THE
ATLAS OF
DISAPPEARING
PLACES

THE ATLAS OF DISAPPEARING PLACES

< < < < < < < < < < < < < <

OUR COASTS AND OCEANS IN THE CLIMATE CRISIS

Christina Conklin *and*
Marina Psaros

MAPS AND ILLUSTRATIONS by Christina Conklin

THE
NEW
PRESS

NEW YORK
LONDON

Support for this book is, in part, from Furthermore grants in publishing, a program of the J.M. Kaplan Fund.

Published in the United States by The New Press, New York, 2021
Distributed by Two Rivers Distribution

LIBRARY OF CONGRESS CATALOGING-IN-PUBLICATION DATA
Names: Conklin, Christina, author. | Psaros, Marina, author.
Title: The atlas of disappearing places : our coasts and oceans in the
  climate crisis / Christina Conklin and Marina Psaros ; maps and
  illustrations by Christina Conklin.
Description: New York : The New Press, 2021. | Includes bibliographical
  references and index. | Summary: "A heavily illustrated book and
  narrative about the threat of rising sea levels around the world"--
  Provided by publisher.
Identifiers: LCCN 2020041259 | ISBN 9781620974568 (hardcover) | ISBN
  9781620974575 (ebook)
Subjects: LCSH: Sea level. | Ocean temperature. | Coastal ecosystem health.
  | Coast changes. | Marine ecosystem health. | Marine pollution.
Classification: LCC GC89 .C66 2021 | DDC 551.45/7--dc23

The New Press publishes books that promote and enrich public discussion and understanding of the issues vital to our democracy and to a more equitable world. These books are made possible by the enthusiasm of our readers; the support of a committed group of donors, large and small; the collaboration of our many partners in the independent media and the not-for-profit sector; booksellers, who often hand-sell New Press books; librarians; and above all by our authors.

www.thenewpress.com

Book design and composition by Lovedog Studio
This book was set in Garamond Premier Pro

Printed in the United States of America

2 4 6 8 10 9 7 5 3 1

To the ocean, and all the people acting bravely to change the story of our fragile, fluid planet.

And to the readers who find this in a used bookstore in fifty years and do what is needed next.

—Christina Conklin

To my father, who gave me a love of the sea. To my mother, who named me after it. And to my children, who will inherit it.

—Marina Psaros

CONTENTS

Foreword   *ix*

Introduction   *xi*

PART I: CHANGING CHEMISTRY   1

KURE ATOLL, HAWAI'I:
PLASTIC, PLASTIC EVERYWHERE   6

THE ARABIAN SEA: REGIME SHIFT   14

CAMDEN, MAINE:
SALT, FAT, ACID, (NO) MEAT   22

THE COOK ISLANDS: FEEDING THE FEVER   28

SAN FRANCISCO BAY: SEVEN-LAYER DIP   34

PART II: STRENGTHENING STORMS   41

HOUSTON: WE HAVE A PROBLEM   46

HAMBURG, GERMANY: RIVER CITY AT RISK   54

NEW YORK, NEW YORK: CAPITAL OF CAPITAL   60

SAN JUAN, PUERTO RICO:
PODER, DESPACITO   68

KUTUPALONG CAMP, BANGLADESH:
HUMAN TIDES   74

PART III: WARMING WATERS 85

THE ARCTIC OCEAN: WHEN THE ICE MELTS 90

PISCO, PERU: ENSO AND THE END OF FISH 98

THE NORTH ATLANTIC: IN DEEP 106

KISITE, KENYA: CORAL COLLAPSE 114

PINE ISLAND GLACIER:
WHAT HAPPENS IN ANTARCTICA
DOESN'T STAY IN ANTARCTICA 122

PART IV: RISING SEAS 129

SHANGHAI, CHINA: SINK, SANK, SUNK 134

HAMPTON ROADS, VIRGINIA:
BYE, BYE, BIRDIES 142

BEN TRE, VIETNAM:
DOING MORE WITH LESS 150

THE THAMES ESTUARY, BRITAIN:
FROM GRAVESEND TO ALLHALLOWS 158

ISE, JAPAN: TRADITION FOR THE FUTURE 166

What's Next? by Marina Psaros 173

Toward Transilience by Christina Conklin 176

Acknowledgments 179

Notes 181

Image Sources 207

Index 210

FOREWORD

Climate change is causing treasured coastal places and marine resources to disappear. Storms have washed away a great many coastal properties and continue to erode beaches and shorelines. Warming waters have melted ice in the Arctic and the Antarctic, releasing the most potent greenhouse gases of all and threatening coastal communities with rising sea levels. Changing water chemistry is destroying crucial marine habitats, altering the balance of aquatic life, and creating threats and challenges to human health. Rising seas (and increasingly intense storms) are making it impossible to sustain public infrastructure and forcing communities to abandon long-cherished areas and historical sites.

Marina Psaros and Christina Conklin provide twenty stories of the impacts that these climate-driven changes are having all over the world. They present their work in the form of a visually engaging and creative atlas, but their goal is to pinpoint the serious and difficult choices facing governments, companies, communities, and individuals. We have to decide whether we are going to do anything to protect ourselves and our planet, and if so, what?

Climate mitigation refers to the ways we can reduce greenhouse gas emissions and halt the long-term warming of the planet. Climate adaptation focuses on the means by which communities can blunt a range of climate change impacts—which are already inevitable—and enhance local resilience. People all over the world need to wrestle with both. The obvious adaptive responses in coastal areas are to waterproof buildings, build sea walls, "harden" critical infrastructure like water treatment and waste disposal plants (so they can withstand increased flooding), and rely more on "green" buffers and wetlands to absorb coastal impacts. The less appealing measures are to abandon certain areas altogether by buying out landowners and explaining to them that public services can no longer be rebuilt each time they are destroyed by floods or storms. There are small island nations around the world that have no choice but to pick up now and move to inland locations, abandoning their history, culture, and property. The choices are painful.

While individuals need to educate themselves about their resiliency, the majority of adaptation decisions need to be made collectively. That is, whole communities must decide which trade-offs they want to make and how they want to use their public resources. It is unlikely that state/provincial or local governments are going to force individual landowners to abandon their coastal or riverine properties, but they may have no choice but to tell those same viruses in every liter of seawater continually co-create the ocean's ever-changing body. Since we are most familiar with our own fleshy vessels—how these bags of salty water feel and how they function—we explore the parallels between our human bodies and the body of the ocean. Each of the four sections of the book examines analogous systems in humans and oceans so that we can more closely identify with—and possibly empathize with—the ocean, our original home.

We explore these parallels in each of the book's four parts. In Part I, we confront our problematic behaviors that are the root cause of climate change. We show how our dependence on fossil fuels is changing the biology and even the basic chemistry of the ocean in the same way that substance abuse changes the biology and chemistry of the user. We are abusing chemicals which are, in turn, flowing to the sea and ruining coastal ecosystems and drinking water supplies. We are killing marine animals with plastic pollution and ourselves with chemical contaminants. Growing ocean dead zones and species extinctions are signs of mounting trauma; the distress signals from a situation that has gotten out of hand.

Part II tells stories about the strengthening storms that are already traumatizing ecosystems and communities, often hurting already vulnerable people and places the most. Doctors refer to storms that sweep through human bodies, from a storm of cholera surging through a community to a depressive storm that rages beyond the control of the sufferer.[5] Traumatic events change us—just as hurricanes and cyclones change cityscapes and coastal habitats—but recovery is possible, especially for those who have worked on their resilience in advance. The way to most fully heal from both personal and societal trauma is to identify the root causes, address them, and then move forward, scars and all.

In Part III, we compare the ongoing systemic stress of ocean warming to a fever. In our bodies, fever is protective to a point, though too much can cause significant, even permanent, damage. In the ocean, increasing water temperatures (1 to 4 degrees Celsius, or 1.8 to 7.2 degrees Fahrenheit, on average this century, with higher spikes in some places) are already throwing food webs out of balance, with unknown consequences. Higher temperatures are also causing ocean waters to circulate more slowly, which could have major impacts on weather patterns. Warmer water takes up more space, contributing significantly to sea-level rise. Heat is also starting to trigger dangerous feedback loops at the planet's poles. Like us, the ocean has the ability to self-regulate within limits; but climate change may cause a runaway fever which could shut down the earth's essential organs, just as our essential organs can shut down when they overheat.

By the time significant rising waters arrive, the subject of Part IV, our present dis-ease will have become a critical disease, and therefore much harder to treat. Even though we could have addressed it earlier, these late-stage systemic symptoms will show up everywhere, all at once. Treatment will be extremely expensive—bankruptingly so—but as with emergency heart surgery, we may have no choice at that point. Consumption, the old-fashioned name for tuberculosis, described the condition of drowning in one's own fluids. Consumption is an apt description of our contemporary malady, too.

There are many different ways to explore this Atlas. Some readers will dive straight in to the story of a particular place, others will want to proceed section by section for a com-prehensive overview of each issue. Reading the book from beginning to end offers an allegory of health and illness, challenge and renewal. Whatever your path, we hope that you find inspiration for how we can collectively write the next chapter in the history of our home.

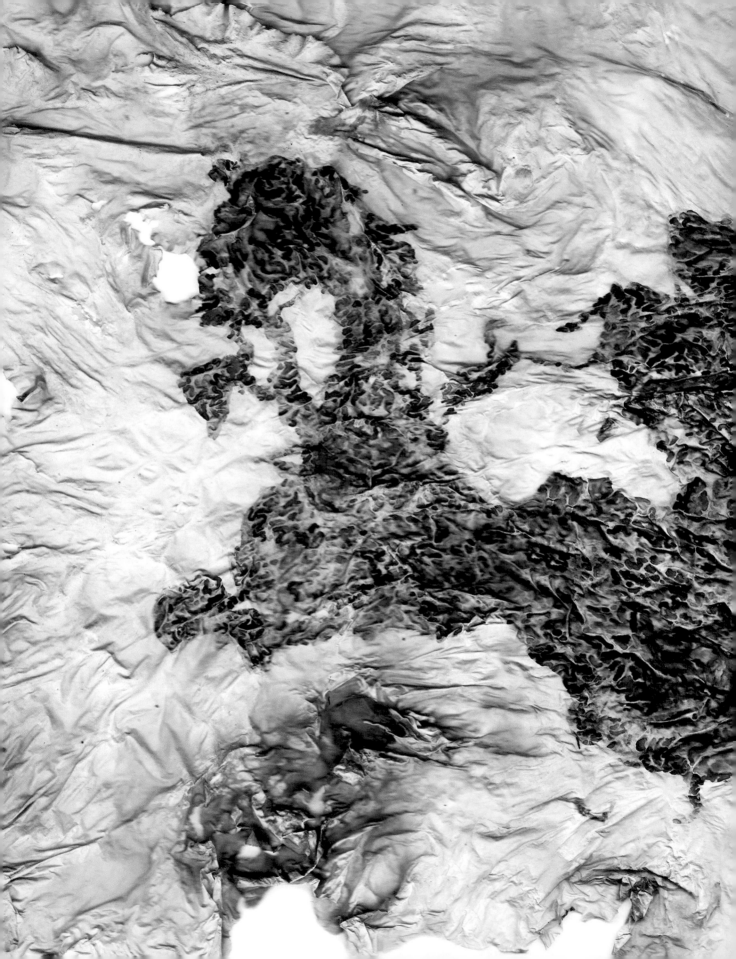

The Spilhaus Projection shows the ocean as a single body of water. First published in 1979 by South African geophysicist and oceanographer Athelstan Spilhaus, it places the earth's axis through China and Argentina, allowing the water to be contiguous and the land to be distorted. The only cut is at the Bering Strait, a narrow channel between Russia and North America, which becomes exposed land during glacial periods.

Part I

CHANGING CHEMISTRY

The water in our cells is the same as the water at the bottom of the ocean. No matter how you look at it, it's still two hydrogen atoms attached to an oxygen atom. I love the kinship that chemistry implies.

—Robin Wall Kimmerer, botanist and writer

"The ocean is a vast chemical solution," writes French geochemist Catherine Jeandel.[1] Salts—compounds that result from the reaction of an acid with a base—make up around 3.5 percent of the weight of the ocean, with just five chemicals making up 99 percent of these salts: sodium, chloride, magnesium, sulphate, and calcium. The remaining 1 percent is comprised of trace elements like potassium, iron, bromine, and carbon, which are essential to life in tiny quantities. Iron, for example, is found in only one part per billion of seawater, yet it is required for photosynthesis. When living things die, they return to their mineral forms after being partially or fully digested by bacteria, creating the life cycle that links animals, vegetables, and minerals into the biogeochemical loop.[2] Even the tangy, briny smell of the ocean is the result of chemical compounds including dimethyl sulfide, which is created by the bacterial consumption of decaying plankton, and bromophenols produced by marine algae.[3]

Every action in the ocean triggers a chemical reaction that cascades through its systems, and most of the sea's current crises hail directly from human pollution. A century of added plastics, petrochemicals, and excess carbon dioxide has fundamentally altered the eons-old chemical composition of the ocean.

Nitrogen- and phosphorus-based fertilizers, which are often over-applied by farmers, flow into rivers, and those runoff-filled rivers flow to the sea. The excess nutrients mix with warmer water, permitting more coastal algae to thrive. This leads to toxic algal blooms that consume all the oxygen in the water, suffocate all that lives beneath, and poison fish and the people who eat them with a range of debilitating illnesses. Chemicals from shoreline dumps, industrial sites, military bases, and airports leach into the surrounding environment, where they poison fish, birds, drinking water, and the air we breathe. Plastics, which are direct derivatives of petroleum processing, are now found from the Arctic to the Antarctic, in both the deepest marine trenches and the food we eat, though their impact on our health remains unknown.

Fossil fuels are, ironically, mostly made up of the compressed bodies of tiny sea creatures that were buried deep underground millions of years ago. Burning their carbon sequestering bodies is causing the ocean to become more acidic, because salt water reacts with increasing atmospheric carbon to form more carbonic acid in the sea. The ocean has absorbed around 26 percent of the carbon dioxide we have released over the past few centuries.[4] Though this has saved us from even warmer air temperatures, the ocean has rapidly paid the price, with pH levels changing more in the past three hundred years than in the prior 65 million years.[5] Lastly, warmer waters physically hold less oxygen, so areas that were already "oxygen minimum zones" are now becoming "dead zones." In some places where open ocean dead zones are growing, adjacent coastal dead zones are growing too, leaving fish, sharks, whales, and marine invertebrates with less and less suitable habitat.

Since the ocean is slow to absorb our skyrocketing rates of atmospheric carbon, the pace of acidification and warming lags behind our emissions by twenty-five to fifty years at the surface, and longer in the deep sea. This lag allows us a false sense of confidence regarding the manageability of global heating, but our current and historical pollution will continue to catch up with us for decades, even after we have reduced global carbon emissions. The sobering reality is that these trends in warming, acidification, and deoxygenation aren't reversible in our lifetimes, or even in our grandchildren's lifetimes.[6]

Getting It Out of Our Systems

Like the ocean's ecosystems, our own systems require hydrogen, oxygen, sodium, and nitrogen, as well as micronutrients like iron and iodine, to maintain a healthy chemical balance. When this equilibrium is upset by behavioral, environmental, or genetic stressors, ranging from excessive drinking to toxins in our water supply to a predisposition to certain diseases, our bodily systems can be overwhelmed. We may only realize we are very sick once the symptoms have become unmanageable, even though our blood, kidneys, and liver have been working overtime for years. In chronic diseases from diabetes to addiction, if these organs fail, we know our time is short. Intensive treatment can lead to significant healing, but without addressing

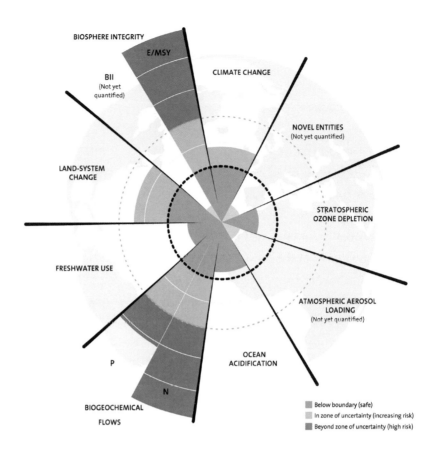

The Stockholm Resilience Center proposes nine planetary boundaries that indicate global system health. Biosphere Integrity and Biogeochemical Flows, namely nitrogen and phosphorous levels, are already "beyond the zone of uncertainty."

the root causes of illness, the disease will simply return, perhaps stronger than before. We have been polluting our ocean too much for too long, and this is rapidly leading to potential long-term disability.

The root cause of our planetary disease is our addiction to fossil fuels. Our reliance on oil and gas, plastic containers, synthetic fertilizers, microfibers, and the other trappings of modern convenience culture are getting us in trouble, and the devastating side effects will only begin to heal after we've stopped using. Our recovery first requires that we recognize this dependence and the system of habits, consumption, and lifestyle choices that fuel it. The next step is to make conscious decisions that put us on a different path, one that is

based on what we want for the future rather than who we were in the past.

Recent discoveries in medicine, psychology, and public health have changed how we think about addiction, dependence, and recovery. This shift allows people in recovery to challenge the system and their place within it, and to identify what resources they need to make healthier choices. Approaches like harm reduction focus on making incremental changes that reduce the negative consequences of use, rather than insisting on a perhaps unattainable (or unattainable right now) abstinent perfection. For example, even cutting out one plastic item from our lives, like single-use plastic water bottles, reduces harm. Such individual steps can be one small part of the restructuring of consumption patterns that is required to save future generations from catastrophic consequences. We also know that we must collectively do more, since time is short to address the serious state of our disease.

One working theory in addiction science is that people with substance abuse issues stop maturing once they start abusing; they become frozen at the emotional age that they began buffering and stopped learning.[7] By this model, most of the industrialized world is stuck in the mid-twentieth century, when post–World War II capitalism boomed along with populations, proliferating cars and chemicals, pollutants and pesticides. It's time we grow up and join the twenty-first century.

The first step in nearly every recovery approach is to recognize that the old behaviors aren't working anymore. Self-deceit is inherent in addictive systems—we've been living in unsustainable ways for decades and haven't confronted it head on. But self-examination and accountability can make it possible to move toward a life of integrity and dignity. Creating and holding on to a vision of a better life ahead can motivate change, as can rejecting the blaming and shaming about how we got into this mess and instead starting to problem solve for how we get out of it. Some changes we can make as we confront our addiction will be relatively easy, like using a bike instead of a car for light errands or opting for the veggie burger instead of beef. Some will need to be a combination of personal and structural change; we can each cut our own plastic use, for instance, but if the plastic is still being produced in mass quantities and used in every school cafeteria, coffee shop,

and grocery store, then impactful change isn't happening. Other solutions will require massive international political commitment, like creating binding agreements to significantly reduce greenhouse gas emissions. It will take time and a lasting commitment to equitable solutions, because it is neither feasible nor fair to immediately eliminate all nitrogen-based fertilizers, tarps, plastic jugs, and oil cookstoves from the daily lives of billions of people. Joint action by the global community of regulators, businesses, and citizens has to happen in order to solve the problem. It will be hard but not impossible.

The news sounds bleak, much like the news about acid rain did in the 1980s. But the global community came together quickly in response to that threat, agreeing on the Montreal Protocol in 1987, which phased out many dangerous ozone-depleting substances. It was the world's first universally ratified international treaty, and it has been repeatedly updated to reflect new science. Unfortunately, when it banned chlorofluorocarbons (CFCs), industries replaced them with hydrofluorocarbons (HFCs)—which don't negatively affect ozone levels, but which *are* very potent greenhouse gases. So in 2017, the international community came together again to add the Kigali Amendment, a binding agreement to reduce HFCs by 80 percent by 2049. It has been ratified by more than one hundred countries so far, with more expected, providing a critical opportunity for countries to come together to address global warming. The United States is not yet a signatory, although the European Union, the United Kingdom, Japan, and Canada are.[8]

In the United States, the 1990 Clean Air

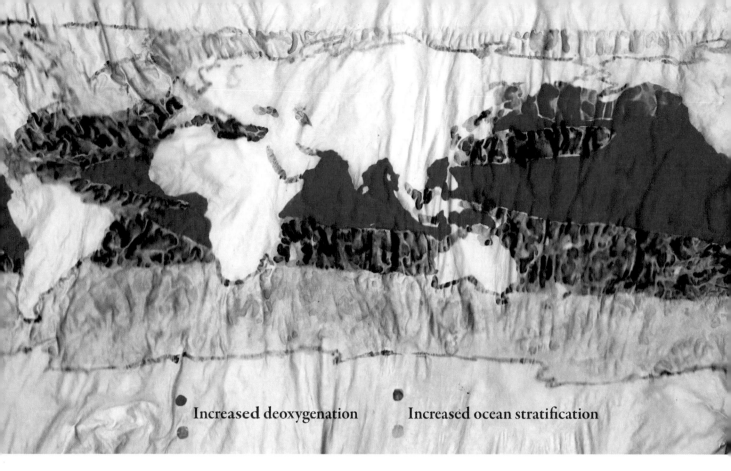

Increased deoxygenation   **Increased ocean stratification**

Increased deoxygenation and stratification in the ocean are permanently changing its basic chemistry.

Act Amendment devised a cap and trade system to reduce the pollutants causing acid rain, proving that industries could quickly adapt and that markets would not fail. In this instance, regulation forced technological innovation and was successful more quickly than anyone predicted. To be clear, cap and trade programs for carbon emissions don't go far enough to address climate change. Even if it is fully implemented, the targets agreed to by signatories of the 2015 Paris Climate Accord will likely lead to 3°C of global warming, well beyond the "safe" boundary agreed upon by Intergovernmental Panel on Climate Change (IPCC) scientists of 1.5°C. Governments around the world must make ever-more robust regulatory agreements that protect citizens and ecosystems from those who would keep us hooked on fossil fuels and sell out our children's future.

Russia

Alaska

Covering almost one million square kilometers (the area of Mongolia or Iran),
the North Pacific gyre contains roughly
1.8 trillion pieces of plastic. In many East
Asian countries, 75 to 90 percent of plastic washes into the ocean. In the United
States, only 2 percent of plastics do, but
consumption is so high that the U.S. is the
twentieth worst ocean polluting country in
the world.

Kure
Atoll

North Pacific gyre or
"Great Pacific Garbage Patch"

Papahānaumokuākea
National Marine
Sanctuary

Kilograms of plastic per
square kilometer

100

10

1

.1

Hawai'i

KURE ATOLL, HAWAI'I: PLASTIC, PLASTIC EVERYWHERE

A plastic bag isn't just for humans, it's for seagulls to choke on. . . . A Styrofoam cup isn't just for coffee, it's for being slowly digested by soil bacteria for 500 years. A nuclear device isn't just for your enemy, it's for beings 24,000 years from now.

—Timothy Morton, *Being Ecological*[1]

This is the story of a T-shirt. It began its life as crude oil pumped in Nigeria, Libya, the United States, and Iran, and decades later it ended the life of a leatherback turtle on Kure Atoll in Hawai'i. This is the story of the plastic we wear and what happens to it.[2]

Kure Atoll sits smack in the middle of the northern Pacific Ocean, fourteen hundred miles northwest of Honolulu and almost exactly halfway between San Francisco and Tokyo. The westernmost islet of the Hawaiian chain, it has been awarded many layers of bureaucratic protection, funding, and recognition. It is a State Seabird Sanctuary; part of the Papahānaumokuākea Marine National Monument; a UNESCO World Heritage site; and an internationally recognized Particularly Sensitive Sea Area. But despite all these designations and the atoll's remote location, Kure's beaches are littered with the plastic-filled carcasses of sea birds and other creatures that depend on this tiny shard of sand as a resting place and breeding ground.

One leatherback turtle—a two-meter-long female from Zihautenejo, Mexico, who first swam to sea fifty-two years ago—had eaten many bits of plastic bags over the years, thinking they were jellyfish. She had also swallowed plenty of polyethylene beads—the base stock of plastics manufacturing—which she had mistaken for fish eggs, as well as the

synonymous with "tipping points," can be difficult to determine because of the complex systems in which they are located. Yet they are critical to understanding environmental systems because crossing them leads to irreversible change. Among the Stockholm Resilience Center's nine planetary thresholds that imperil life on earth, we have already exceeded the safe bounds of four systems— biosphere integrity, biogeochemical flows, land-use, and climate change. The thresholds for "release of novel entities into the environment" and atmospheric aerosol loading are systems whose boundaries are not yet even defined, much less accurately measured.[10]

One of the most alarming regulatory failures in the United States is that most pollution rules control a single contaminant, not larger classes of related pollutants that pose similar health risks. This allows companies to switch to unregulated alternatives among the tens of thousands of chemicals on the market that have not yet been proven dangerous. Using a more cautious approach, Europe's REACH regulation of "no data, no market" places the onus on companies to collect safety information on the substances they produce, which the European Chemical Agency then disseminates.[11] We know that the monomer building blocks from which polymer plastics are made are carcinogenic, and we know that plastics combine into myriad noxious compounds when they are burned in incinerators or as they break down over time. Will we cross a pollution threshold and not even realize we have done so?

A View from 2050

The Plastics Century (1950–2050), which took flight with the Baby Boom and soared for decades, has played out in some surprising ways in its second half. In the early decades of the twenty-first century, scientists made the connection between petrochemical production, the pervasive use of plastics, and rising cancer rates, but it took many years for regulators and the public to take these health impacts seriously.[12] In the 2010s, as photos of garbage-strewn beaches and strangled seals were joined by ominous headlines about microplastics in the air, drinking water, soil, and ocean, consumers began to ask more questions about the longer-term impacts of their convenience-driven choices.

In the early 2020s, the American Medical Association and the American Association of Retired People collaborated on a wide-ranging study of plastics and health. They found significant chemical concentrations in the blood of Baby Boomers, who had spent many decades microwaving frozen meals and wearing synthetic clothes.[13] They also learned that most medical devices were made of single-use plastic, despite plastic's health impacts, and that the health-care industry produced nearly a billion pounds of plastic waste a year. Disgusted, they lobbied Congress to regulate low-use consumer plastics as part of the larger 2026 Environmental Regeneration Bill.

Consumers tired of murky eco labels like biodegradable, compostable, all natural, bioplastic, and plant based. Producers had used them to mislead people about the environmental and health properties of their products and packaging. In fact, the green plastics and plant plastics of the early twenty-first century were often full of copolymers, synthetic plasticizers, and other additives that could only be broken down in

special industrial incinerators.[14] When people finally realized all those potatoware forks had been going straight to the landfill for years, voters passed a ballot measure in California that forced accurate labeling by consumer goods companies and levied a stringent "producer pays" tax on all food-related bags, utensils, and packaging. The Plastics Industry Association lobbied hard against it, citing the eighty thousand Californians employed by plastics companies, but lost in the court of public opinion.[15] Many industries nationwide, especially health care and food services, adapted quickly in order to meet California's stricter standards.

At the same time, some Japanese firms were moving toward circular, "cradle-to-cradle" models of manufacturing and designing innovative packaging reuse and recycling initiatives.[16] They scaled a number

Seven out of the top ten sources of ocean pollution are plastics

of eco-technologies the previous generation had invented but never brought to the global market, such as including microfiber filters in all washing machines and building closed-loop polyester recycling plants that broke down fibers into their purest basic units so they could be recycled five to ten times. In 2016 Japanese scientists had also discovered a plastic-consuming bacteria, followed by a fully contained digestion system that could harness the munching microbes in environmentally safe ways.[17] Asian fashion designers launched the ubiquitous "dot tags" now standard on clothes: red dots for recycled, blue dots for refurbished, green dots for compostable, and black dots for new materials. The 2029 fashion shows in Tokyo were dominated by compostable clothing, inspiring designers in Paris and Milan to reinvent the paper dresses of the 1960s with materials made from annual crops, not trees.

Yet, even with advances in materials science and consumer behavior, new plastic production continued to mushroom worldwide. Fast fashion still thrived in many high-income countries; in 2030, the average American adult generated one hundred pounds of textile waste each year, only 15 percent of which was ever reused or recycled.[18] The rest went straight to the landfill. In developing countries, plastics continued to clog waterways on their way to the ocean. The worst plastic polluters, mostly lower- and middle-income countries with large populations and poor solid-waste management systems, worked to upgrade their dumps to twentieth-century standards, but progress was slow.[19]

By 2037, tropical tourist locales had become so choked with garbage that a number of Asian countries agreed to a binding, decade-long cleanup of the South China Sea, a deal that helped keep the peace in those politically fraught waters. The Chinese led the way, developing "last-chance capture" technologies that automatically filtered and incinerated all the macro-trash flowing out of their most polluted rivers, the Yangtze worst among them. This technology alone reduced global ocean pollution by 25 percent.[20] A United Nations program to spread this clean-tech to the most polluting countries around the world was rolled out throughout the 2040s, funded by a voluntary fee the oil majors had agreed to as the cost of continuing to do business. Only then, once the firehose of supply had been slowed to a relative trickle, did ocean pollution slow significantly to just ten million tons per year, less than one-tenth of the "business as usual" rate that had been predicted.[21]

These measures slowed large plastic inputs to the world's rivers and oceans, but no technology could deal with the microplastics that pervaded the earth's ocean, soil, and air. As researchers found ever more epigenetic links between petrochemical exposure and human illness, money poured into treatments for diseases that had been far less common a century before: breast and blood cancers, autoimmune diseases, Type I diabetes, Alzheimer's, and erectile dysfunction.

Consumer demand has now pivoted, driving markets toward renewable fibers, such as spider silk made by bacteria and carbon nanofibers made by converting carbon diox-

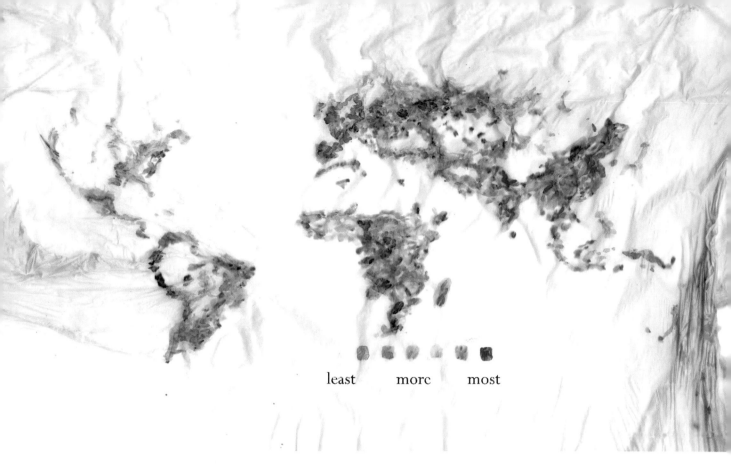

least   more   most

The current annual rate of 60 to 100 million metric tons per year of mismanaged plastic waste is expected to nearly triple by mid-century. Ninety-one percent of this waste washes down rivers to the ocean.

ide into cloth through electrolysis.[22] The tidal wave of plastic production has waned, and well-researched, non-toxic polymers are now used sparingly for products deemed essential like crash helmets and major medical devices.

Pollution levels on Kure and in the North Pacific Gyre have slowly begun to decrease, but researchers expect it will be at least several more generations before the Plastics Century can be put in the past.

THE ARABIAN SEA: REGIME SHIFT

I am the sea. In my depths all treasures dwell.

—Muhammed Hafiz, nineteenth-century Persian poet[1]

In the early 2000s, scientists at Columbia University learned that the food web in the Arabian Sea had suddenly shifted from being based on diatoms, the one-celled algae eaten by zooplankton, to being dominated by a larger, single-celled, luminescent species of dinoflagellate, *Noctiluca scintillans*, which had never been seen in the region.[2] They found that *Noctiluca*, which are omnivores, were thriving in the Arabian Sea's growing "dead zone," or oxygen minimum zone (OMZ), because they could both photosynthesize and feed on the diatoms. At one millimeter in diameter, *Noctiluca* are too big for most zooplankton to eat, so within a decade, *Noctiluca* outcompeted diatoms and had become the base of the Arabian Sea's food chain. Then the zooplankton dwindled and fish populations crashed, while jellyfish, sea salps, and turtles feasted on *Noctiluca*, throwing the whole ecosystem into flux. "Something dramatic has changed in the Arabian Sea," according to Andrew Juhl, a microbiologist at Lamont-Doherty Laboratory at Columbia University.[3] The largest and thickest dead zone in the world now swirls at its heart.[4]

Though it looks homogenous, the open ocean is actually a dynamic patchwork of loosely woven layers that mix with each day and each season. Some layers with specific temperature and chemistry are just a few centimeters thick, while others run deep and wide. Each of these liquid layers is essentially stable, with its own consistent chemical "weather." Every marine species has evolved to suit its particular microclimate of oxygen saturation and light. It is at the busy

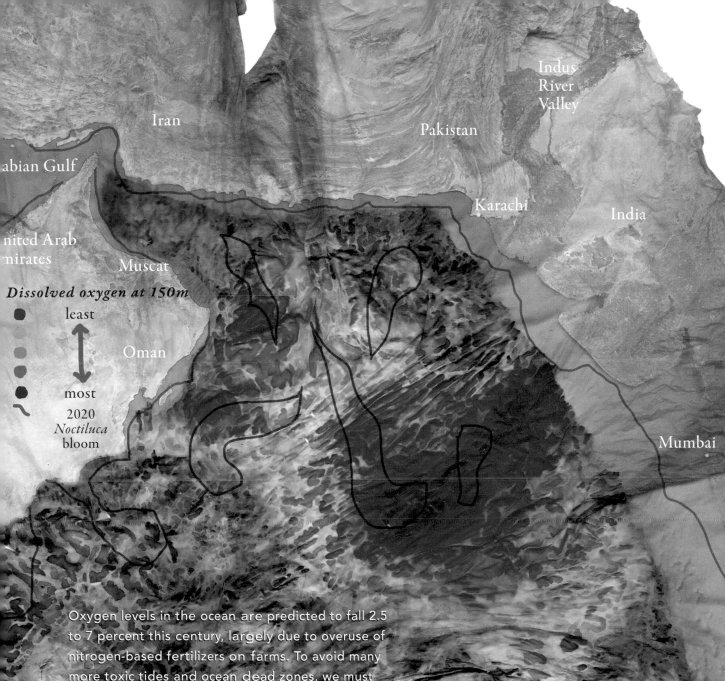

Iran

Indus
River
Valley

Pakistan

abian Gulf

Karachi

India

nited Arab
nirates

Muscat

Dissolved oxygen at 150m

least

↕

most

Oman

2020
Noctiluca
bloom

Mumbai

Oxygen levels in the ocean are predicted to fall 2.5
to 7 percent this century, largely due to overuse of
nitrogen-based fertilizers on farms. To avoid many
more toxic tides and ocean dead zones, we must
double fertilizer efficiency and halve nitrogen waste.
In the Arabian Sea, the problem is exacerbated by
Himalayan glacial melt and warmer monsoon winds,
which reduce photosynthesis in the surface layer and
give *Noctiluca* an advantage over diatoms.

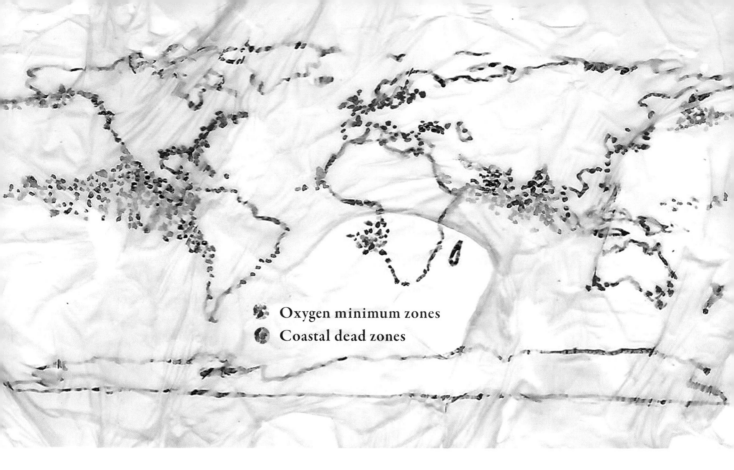

Oxygen minimum zones
Coastal dead zones

Since 1950, more than 450 new coastal dead zones have been identified, and in some of these places regime shift is a major risk.

intersections between the strata where the food web is most diverse, enlivened by the exchange of nutrients and creatures.

Each June, as the southwest monsoon blows off the Himalayas toward Africa, an enormous upwelling of nutrients rises from the deep sediments along the coast of Oman and causes an explosion of life. Then, in February, the winds reverse, and mangrove forests along the shores of southern Pakistan and western India create a safe, abundant breeding ground for many sea-going species. This large marine ecosystem used to be replete with life. Dugongs (or sea cows), turtles, dolphins, yellowfin tuna, sharks,

and other marine royalty filled the waters surrounding its central OMZ. Such naturally occurring OMZs in the open ocean are formed by typical water circulation patterns and cause no harm.

Yet over the past fifty years, because the seas have warmed and warmer waters release oxygen molecules more readily, average oxygen levels have dropped by 2 percent, or 77 billion metric tons. In the central Arabian Sea, this shift is causing already low biodiversity to decrease and allowing bacteria, viruses, and protists like *Noctiluca* to thrive. Around the world, ocean that is completely devoid of oxygen has more than quadrupled in recent decades, growing by an area nearly the size of the European Union.[5] In the Arabian Sea, Bastien Queste, an oceanographer who sent remote-control submarines into this poorly studied region, recently assessed that "the situation is actually worse than feared. . . . The ocean is suffocating."[6]

Warmer water also stratifies more readily and is more resistant to mixing, which accounts for half of the oxygen loss in the top one thousand meters of the ocean. However, living things in warmer environments need *more* oxygen to fuel their higher rates of respiration and metabolism. So creatures with finely tuned oxygen tolerances must either migrate, suffocate, or experience often permanent hormone and neurological changes. Swimmers like fish larvae can survive for a time in an OMZ, but their oxygen-starved bodies make easier prey. Fish kills, where millions of fish simultaneously die and wash up on shore, have increased dramatically in the region over recent decades. Fewer fish means fewer meals for major predators, so emaciated whales and dolphins have been washing up on Indian beaches with greater frequency.[7]

When large predators must venture closer to shore to feed, small-boat fishermen rush to catch them. In 2014 alone, more than 213,000 tons of large sharks and 41,000 tons of large rays were killed for their fins and gill plates, which are popular with Chinese buyers. Half of the shark species in the Arabian Sea are now threatened with extinction—a higher proportion than anywhere else in the world. Giant guitarfish, sawfish, and hammerhead sharks are the region's species in greatest danger, while the Arabian Sea

humpback whale, unique to these waters, has only a few dozen adults remaining.[8]

For fixed or slow-moving animals like corals and crustaceans, there is no escape from ocean deoxygenation. However, jellyfish, salps, bacteria, and protists can thrive in these zones. In the lowest-saturation and oxygen-free waters, only ancient unicellular creatures that metabolize nitrogen can succeed. Yet these anaerobic life-forms emit nitrous oxide rather than oxygen as their waste product; nitrous oxide is a greenhouse gas three hundred times more potent than carbon dioxide. With the ocean predicted to warm four times faster in the next sixty years than in the last sixty years, OMZs are most likely to continue expanding for the forseeable future.[9]

Nearer shore, different problems are disturbing the Arabian Sea's delicate chemical balance. India's biggest city, Mumbai, has doubled in population to 21 million in

Noctiluca scintilans

plankton bloom like exploding fireworks, doubling daily in size in fractals of red, green, or white, depending on the microorganism. Visible from space and often larger than many countries, these blooms block the sun from reaching the life beneath them. As the plankton die, billions of their tiny carcasses rain down through the water, where bacteria digest the dead critters, consuming all the oxygen in the process. Predictable spring and fall algal blooms have always been part of this ecosystem and many others, but the species, scale, and frequency of these events over recent decades are new.

The chemical balance of nitrogen, phosphorus, and oxygen is a determining factor in ocean health. Of the nine thresholds of planetary health defined by the Stockholm Resilience Center, nitrogen loads are already far beyond the "zone of uncertainty (high-risk)."[11] Overfishing, slower monsoon winds, higher sea temperatures, and increased salinity have helped push life in the Arabian Sea onto a knife's edge. According to marine ecologist Denise Breitberg, "once oxygen levels are low, behavioral and biogeochemical feedbacks can hinder a return to higher-oxygen conditions." This is called a regime shift, ecologists' term for an irreversible change from one state to another, "even when the driver that precipitated the shift

the past decade, dumping vast amounts of untreated sewage into the sea. The Indus River, which flows down the spine of Pakistan, flushes thirty million gallons of sewage into coastal waters every day. But the bigger issue is overapplication of nitrogen- and phosphorus-based fertilizers, which cause eutrophication, or the enrichment of river and coastal waters with millions of tons of excess chemical "nutrients." Fertilizer use has increased tenfold since 1950 as farmers have sought to maximize yields, and millions of lives were saved and improved by the First Green Revolution (1966–1985). Now, though, farmers have become dependent on agricultural chemicals that we know are being oversold and overused. In India, 70 percent of coastal pollution comes from fertilizer, and 80 percent of fertilizers used wash into the sea.[10]

When these chemicals reach the sea, they fan out to form a massive food source for phytoplankton, which use these naturally rare elements to photosynthesize. The is reduced or removed." These systems never bounce back.[12]

Key Term: Precautionary Principle

Vorsorge, a German word meaning precaution with an emphasis on taking preparatory action, is a central idea in European environmental policy used to determine "actions when weakly understood causes could lead to catastrophic or irreversible events."[13] Precautionary frameworks have four elements: 1) taking preventive action despite uncertainty; 2) shifting the burden of proof to the proponents of an action (for example, requiring chemical companies to prove the safety of their products before introducing them); 3) exploring a wide range of alternatives to avoid possibly harmful actions; and 4) increasing public participation in decision-making. Bringing these criteria into wider use can protect countless species and ecosystems.

The World Health Organization notes that, in the past, we thought removing hazards (environmental toxins, for instance) in reaction to a proven threat led to improved public health. Now it is understood that precaution and prevention are important steps that should be taken before the need to react ever arises. These principles also encourage innovation and cross-disciplinary problem-solving in ways that reduce harm and save money in the long run. There is a "complicated feedback relation between the discoveries of science and the setting of policy," according to environmental scientist David Kriebel and public health scientist Joel Tickner. They urge their colleagues to prioritize research that protects people and the environment and to remain aware of the potential policy implications of their work.[14]

A View from 2050

By 2025, 80 percent of fish stocks in the Arabian Sea had been overexploited or had collapsed completely, and coastal fishermen throughout the region, already marginal in every sense of the word, faced hunger for the first time in two generations. In response, Sultan Haitham bin Tarik of Oman, a country with a long tradition of peace-making and non-alignment, convened a regional conference to address climate and ocean health.[15] The event's major announcement was an agreement to increase renewable energy production, improve agricultural practices, and enforce stricter fish-catch limits, commitments that were already aligned with each nation's goals. More daring was the decision to develop joint strategy documents and share scientists between countries with differing political allegiances. Smiling kings and presidents gathered for group photos, congratulating themselves on a genuine step forward. In a side deal, Oman brought together mortal enemies Saudi Arabia and Iran to work together on large-scale algae farms—long, serpentine tanks of phytoplankton called "raceways"—that would generate both biofuel and export revenue. By 2030, the Arabian deserts bloomed with algae, a successful pilot that led to further joint investment, inching the region toward greater stability.

Sultan Haitham also began funding Omani biologists to develop more heat-tolerant giant kelp species and establish large offshore kelp farms along the country's long, barren coast. Reforesting the sea at a huge scale had already proven successful in the Atlantic, where the kelp sequestered carbon dioxide from the air and helped reestablish the failing coastal food pyramid. In Oman, kelp helped eventually bring back both small-scale fishing and shark populations.[16] In his 2031 interview on the project's success, Sultan Haitham said, "Human intervention can indeed be a force for good, benefitting wildlife, biodiversity, water quality, global warming, people, and economies."

On the other side of the ever-expanding dead zone, India and Pakistan struggled to reduce fertilizer use and improve waste management. In 2028, after several years of hunger due to recurring algal blooms, a coalition of farmers and fishermen in Kerala brought a complaint to the National Green Tribunal, the country's independent court for environmental justice issues. In its early years, the tribunal had nearly become another bloated and toothless bureaucracy. But the crisis in the Arabian Sea compelled it, for the first time in international law, to legally link land, river, air, and ocean as one cohesive system—a system requiring a comprehensive action plan. It established

A Partial List of Endangered Shark Species in the Arabian Region

Critically Endangered

Stripenose Guitarfish	*Acroteriobatus variegatus*
Pakistan Whipray	*Maculabatis arabica*
Red Sea Torpedo	*Torpedo suessi*

Endangered

Ocellate Eagle Ray	*Aetomylaeus milvus*
Smoothtooth Blacktip Shark	*Carcharhinus leiodon*
Aden Torpedo	*Torpedo adenensis*

Vulnerable

Halavi Guitarfish	*Glaucostegus halavi*
Speckled Catshark	*Halaelurus boesemani*

Near Threatened

Salalah Guitarfish	*Acroteriobatus salalah*
Scaly Whipray	*Brevitrygon walga*
Arabian Carpetshark	*Chiloscyllium arabicum*
Cowtail Ray	*Pastinachus sephen*
Spotted Guitarfish	*Rhinobatos punctifer*

Data Deficient

Oman Guitarfish	*Acroteriobatus omanensis*
Reverse Skate	*Amblyraja reversa*
Arabian Catshark	*Bythaelurus alcockii*
Quagga Catshark	*Halaelurus quagga*
Oman Bullhead Shark	*Heterodontus omanensis*
Eilat Electric Ray	*Heteronarce bentuviai*
Soft Electric Ray	*Heteronarce mollis*
Bigeye Numbfish	*Narcine oculifera*
Pita Skate	*Raja pita*

an Environmental Reconciliation Commission to collect steep fines from offending state governments and corporations, monitor pollution levels, and oversee coastal restoration projects. These projects created jobs for unemployed fishermen and farmer education campaigns that, over two decades, reduced fertilizer use by 40 percent with no decrease in productivity. The tribunal's decision was challenged by global agribusinesses but upheld by the Indian Supreme

Court, and by 2045, healthier soil and fisheries were returning. India is currently leading the way in South Asian coastal zone management.

Pakistan, on the other hand, failed to act. Though a few weak environmental laws were passed around 2000, implementation utterly failed, so Pakistan's coastal dead zone merged with the Arabian Sea's central dead zone, leaving only green sludge for much of every year. Though the prime minister had created a Committee on Climate Change in 2018 to "periodically monitor" adaptation and mitigation efforts, no new laws had been passed by 2030, surprising no one. As the world's seventh most vulnerable country to heat, floods, droughts, and internal climate migration, Pakistan had chosen to focus only on its food supply. Now, in 2050, Pakistan finds itself isolated, a country that needs friends but has few.

The Gulf of Maine is one of the fastest-warming ecosystems in the world, but most attempts to transplant the American lobster, *Homarus americanus*, to other cold-water sites have failed.

United States

Canada

Camden, Maine

Nova Scoti

Massachusetts

New York

Lobster Catch

1967

2014

CAMDEN, MAINE: SALT, FAT, ACID, (NO) MEAT

> For what is the environmental crisis if not a crisis of the way we live? The Big Problem is nothing more or less than the sum total of countless little everyday choices.
>
> —Michael Pollan, *Cooked*[1]

Coastal Maine is glorious in summer, the harbor in Camden full of pretty sailboats taking vacationers out to noodle around, gaze at the stars, swim in the water, and enjoy a lobster dinner on an uninhabited island. At least once a week, crews from the windjammers (traditional sailboats with no engines) delight guests by tipping enormous pots of boiling water and bright red lobsters onto beds of seaweed for a summer feast—a meal that is one of the many pleasures that may disappear as a result of climate change.

Ocean waters are acidifying in higher latitudes more quickly than they are in the tropics, and Gulf of Maine waters are acidifying faster than 99 percent of the rest of the planet.

THE COOK ISLANDS: FEEDING THE FEVER

> What is it that is truly valuable to us? Do we want to be extremely wealthy and have no fish to eat? Or do we want to have enough and live in a clean environment?
>
> —Jacqueline Evans, marine biologist and environmental activist[1]

A fifty-year-old man was admitted to Rarotonga Hospital with heavy diarrhea, vomiting, weakness, and severe joint and muscle pain. Anything cold felt burning hot to the touch, and anything hot was ice cold. Two hours earlier, he had eaten red snapper, or *anga-mea*, a common reef fish, and had unwittingly caught ciguatera, a severe form of food poisoning. The man was given IV fluids and antiemetics, but no further treatment was available, so after three weeks of slow improvement, he was released from the hospital. In the meantime, all his pigs had died after eating the viscera of that same fish, his cat was paralyzed for weeks but survived, his chickens had to be killed, and his neighbor's dogs died after eating discarded fish parts out of the garbage. The man's joint pain and tingling skin lasted for more than a year, one of the lasting health effects that can include numbness in the arms or legs, cardiovascular damage, and retriggered symptoms upon any future consumption of meat, fish, or alcohol.[2]

As inhabitants of a nation of scattered islands and atolls in the South Pacific, Cook Islanders rely on seafood for both protein and income. Yet up to 10 percent of islanders get ciguatera poisoning every year, a sickness caused by the genus *Gambierdiscus*, single-celled protists the width of a human hair that look like slightly squashed yo-yos. Once rare, ciguatera has reached epidemic rates worldwide, with perhaps 500,000 cases a year, and is now the world's most common form of non-bacterial seafood poisoning.[3] For tourists, ciguatera means the sudden, miserable end of a long-planned vacation. For people living throughout the tropics, outbreaks weaken both public health and food security when local fishing grounds are closed for months or years. Colorless and odorless, *Gambierdiscus* is impervious to freezing and cooking, and it bioaccumulates, making people sicker with each exposure. People are forced to buy expensive imported food or risk it on the reef.

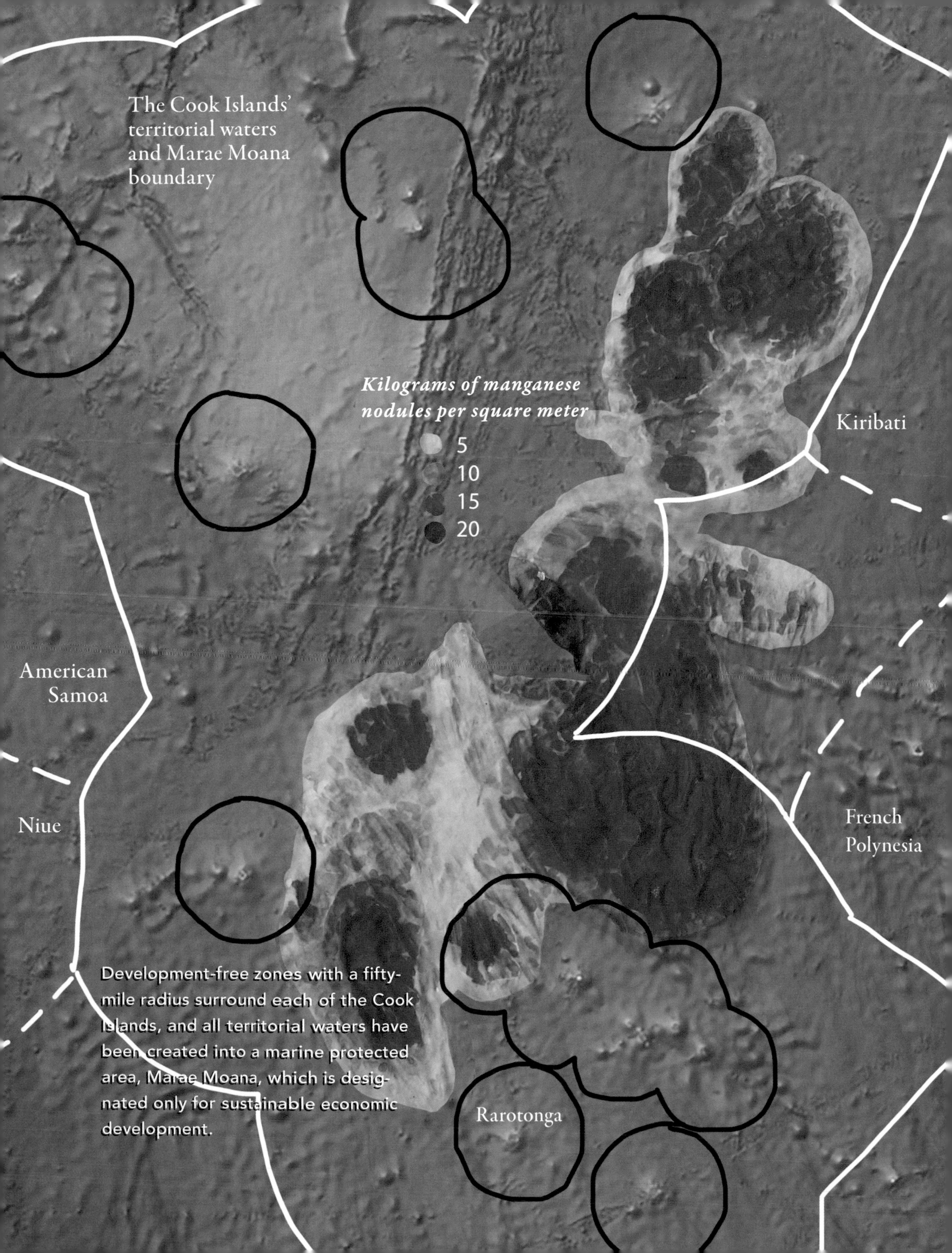

The Cook Islands'
territorial waters
and Marae Moana
boundary

*Kilograms of manganese
nodules per square meter*

5
10
15
20

Kiribati

American
Samoa

Niue

French
Polynesia

Development-free zones with a fifty-
mile radius surround each of the Cook
Islands, and all territorial waters have
been created into a marine protected
area, Marae Moana, which is designated only for sustainable economic
development.

Rarotonga

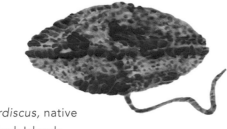

Gambierdiscus, native
to the Cook Islands

Out of five thousand known types of phytoplankton, only eighty produce poisons that cause disease in humans. *Gambierdiscus* dinoflagellates feed on algae that proliferate on coral killed by storms, bleaching, quarrying, dynamiting, dredging, pollution, and landfill. Marine scientists have found that spikes in disease reliably lag two years behind major coral disturbances such as coral mining and the stronger, more frequent cyclone seasons that occur during El Niño years. These days, the fifteen Cook Islands reefs are in relatively good health compared with others, but ciguatera spiked here between 1998 and 2008, causing 2,800 illnesses in a nation of just 15,000 people. Some people are emigrating to New Zealand as island life becomes less secure.[4]

Wealth comes in many forms. In the Cook Islands, wealth is defined less by their 240 square kilometers (93 square miles) of land than by their nearly two million square kilometers (770,000 square miles) of ocean that surrounds and connects them—an area the size of Mexico or Saudi Arabia. Throughout Oceania, people's creation stories, cultural traditions, and economic opportunities hold fast to the sea. But in recent years they have had to balance their deeply held ecological values with the lure of financial wealth the ocean might provide.

After being elected on an environmental platform in 2012, Prime Minister Henry Puna launched a bold vision to convert the country's entire maritime Economic Exclusion Zone (EEZ) into the largest Marine Protected Area (MPA) in the world. "This is our contribution not only to our own well-being but also to humanity's well-being," he said. "The Pacific Ocean is the source of life for us. We are not small Pacific island states. We are large ocean island states."[5] Puna hired Jacqueline Evans, a local marine biologist, to write and shepherd the necessary legislation. Evans drafted a law based on the Great Barrier Reef Act of Australia, which prioritized the protection and conservation of the environment over its economic uses. After determined coalition building, the legislation passed in 2017, dedicating Marae Moana, meaning Sacred Ocean, to ecologically sustainable development. In addition to protecting the country's two-hundred-nautical-mile EEZ, it designated fifty miles around each island as development-free areas, with the aim of restoring struggling reefs and shallow breeding waters. It was a major environmental victory and foreshadowed the United Nations' plan to stabilize global biodiversity by protecting 30 percent of all land and sea by 2030.[6]

Around the same time, a few local businessmen were approached by European mining conglomerates that wanted to explore the possibility of mining manganese nodules from the Cook Islands' sea floor. Manganese, which is required for batteries, steel, fertilizer, soda cans, and ceramics, is the fifth-most-abundant metal in the earth's crust and is widely available around the world. On the sea floor, it primarily occurs as one- to eight-centimeter nodules that have built up over thirty million years. Mining companies want to use "caterpillar" earth-moving machines like those used on land to more easily scrape the nodules off the sea bed. This destroys the deep sea and benthic ecosystems, which are far more diverse than once thought. It also damages the food web, as nutrient-rich sediments are brought through the water column to the surface. One speculative estimate puts the total volume of Cook Islands nodules at 12 billion square tons, with a theoretical present-day value of $10 trillion.[7] Yet deep sea mining is an unproven technology with huge ecological risks in fragile environments.

Immediately after Marae Moana became law in 2017, these local companies began pushing for permission to explore the sea floor in the MPA, something that hadn't been expressly forbidden in the law but was certainly against its intent. Throughout 2018 and 2019, Jacqueline Evans worked to defend Marae Moana's right to exist unperturbed, under the protection of Prime Minister Puna. But Deputy Prime Minister Mark Brown, who has connections to the private concerns, wants to go forward with exploration and then mining. His goal is to create a sovereign wealth fund, like Norway's or Saudi Arabia's oil and gas funds, that will support the country in perpetuity, and he wants to act fast, while prices are strong.[8]

In April 2019, Evans received the world's most prestigious ecological award, the Goldman Environmental Prize, for her work making Marae Moana a reality. In her acceptance speech, she said, "Our people believe the ocean has a personality. We have legends about conversations held with sharks and whales. We have a god of the sea. That is why Prime Minister Puna has called for us to explore the rights of the ocean as a legal entity," a reference to the rights-of-nature movement that seeks to extend legal personhood to natural systems,

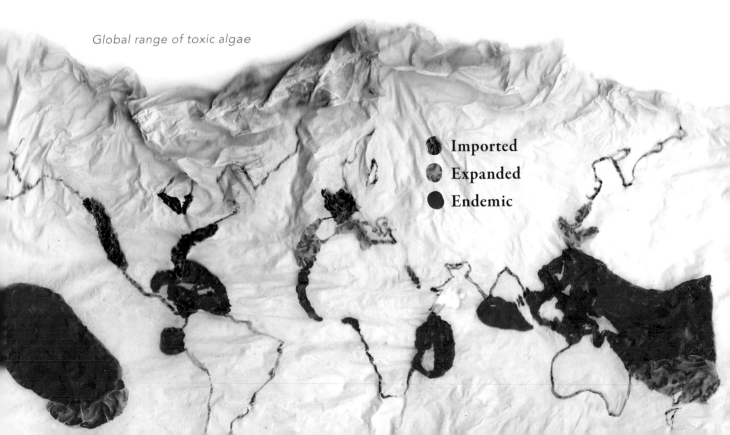

Global range of toxic algae

● **Imported**
◑ **Expanded**
● **Endemic**

the same legal status corporations already enjoy. When Evans returned home with her prize, she advised local environmental groups to call for a ten-year moratorium on seabed mining in Marae Moana, to give scientists time to record baseline ecosystem data that would inform future policy decisions.

She was immediately fired.[9]

~~~~~~~~~~~~~~~~

Key Term: Rights

Thomas Berry, a leading ecological thinker, developed Ten Principles of Jurisprudence that he proposed nations use when considering the concept and granting of *rights*. According to Berry, every component of the earth community has three rights: the right to be, the right to habitat, and the right to fulfill its role in the ever-renewing cycle of life. Rivers, birds, insects, and people all have rights that are role specific and also species specific. Therefore, property rights, for example, are not absolute but relational, since no creature nourishes itself alone.[10] According to the rights-of-nature argument, the ocean, too, has the right to be able to maintain its vital cycles and functions, which are currently being abridged by pollution and anthropogenic climate change. If we decide that both humans and ecosystems have the right to live in healthy conditions, the burden of proof flips: everything is protected by default, and those who want to profit materially must first prove they can do so without causing harm. In 2017, New Zealand granted legal personhood to the Whanganui River, based on rights set out in New Zealand's founding document, the Treaty of Waitangi.

Personhood protects the river from objectification and abuse, and an appointed person represents its interests in all negotiations about its use. Similar legal decisions around the world are being advanced and documented by the Global Alliance for the Rights of Nature.[11]

~~~~~~~~~~~~~~~~

A View from 2050

In the absence of international pressure and effective leadership by the prime minister, the twenty-four-member Cook Islands Parliament passed the Seabed Minerals Act in 2019, which permits exploitation of the sea bed in the name of economic diversification.[12] By 2022, despite pushback from international environmental NGOs and Cook Islanders, the Seabed Minerals Authority began granting mining permits. It was just a few permits at first, but when the money started rolling in, the new prime minister, Mark Brown, opened the floodgates. What had been an opportunity for large-scale environmental protection became another case of an entire economy becoming dependent on a single commodity, a path that has distorted the social and economic policies in many countries for decades.

Using their close legal relationship with New Zealand (which goes back to British colonial boundaries set in 1901), several Cook Island expats in Auckland joined with the Ocean Foundation in 2023 to file a rights-of-nature lawsuit against the government-run Cook Islands Investment Corporation (CIIC). The plaintiffs argued that, as the country's public investment fund, the CIIC must invest in ways that do not abridge the rights of the ocean, citing a scientific report that stated the "impacts of nodule mining in the Pacific Ocean would be extensive, severe, and last for generations, causing essentially irreversible damage."[13] Using the Whanganui River case as precedent, the New Zealand Supreme Court, which had never before intervened so directly in Cook Islands' affairs, determined that their national security obligations to the island nation extended to protecting the environment. "Our environmental laws have literally permitted pollution for too long," the majority opinion declared. "It is our judgement that the meaning of *taonga*—which is translated as 'property' in English but which means 'tangible and intangible treasure' in Maori—extends to natural environments. Permissible uses of the ocean must balance human rights with the rights of other species and the ecosystem itself." They then established an indefinite moratorium on all further mining until it could be proven to be sustainable and safe.[14]

But the ecosystem struggled, as waters continued to warm, cyclones hit more often, and toxic algal blooms, including paralytic shellfish poisoning and diarrhetic shellfish poisoning, came to Cook Islands waters for the first time. The government, always seeking to diversify its economy and broaden its tax base, sent a delegation to Svalbard, the tiny Arctic archipelago north of Norway that had transformed itself into an important international research station in the late twentieth century. When the officials came home—more appreciative than ever of sun and sand—they implemented a plan to build a world-class tropical marine research institute that attracted talent and invest-

ment from all over the world by mid-century.

Largely in response to the climate crisis, South Pacific nations, plus Australia, New Zealand, and Papua New Guinea, began discussing the creation of a political union, not unlike the European Union, with a common currency, free movement, and free trade.[15] It sounded difficult, given the complex territorial claims in the region, but the state of crisis in the region focused minds, as only existential threats can, on cooperation rather than competition. Facing too many migration fires at home, former colonial powers France and Britain ceded their Pacific territories back to local residents. But the United States, unwilling to give up its strategic foothold in the region, only granted American Samoa and Guam observer status. In 2035, after many years of negotiations, the island nations came together in Wellington to sign the Treaty of Oceanic Union, or O.U. for short. It did not solve the problems of widespread poverty, toxic seafood, rising seas, greedy corporate interests, or corrupt politicians, but the countries had greater strength and negotiating power together than alone.

After countless successes—and some setbacks—in her fight for ocean rights around the world, Jacqueline Evans was selected by the member states as the first president of the Oceanic Union, an honorific title since most of the work was done by the secretariat in Fiji. It allowed her to continue speaking on the global stage and to write amicus briefs for supreme courts and environmental tribunals around the world as they decided on the rights of nature. All the times powerful forces had stopped her in her tracks looked, in retrospect, like stepping stones.

Berkeley

Oakland

San Francisco

San Francisco Bay's highly polluted zone is vulnerable to both flooding and sea-level rise, and the infilled land will, in coming decades, be reclaimed by the sea. Most coastal cities share the Bay Area's dual challenge of vulnerable infrastructure along contaminated shorelines.

San Leandro

SFO

Hayward

Fremont

Foster City

Redwood City

Half
Moon
Bay

Blue area = 3 meters flood zone

Palo Alto   Milpit

PFAS surfactants
Landfill
Airport   Mountain View   S
Superfund site   Je
Chrome plating facility
PBDE flame retardants
Unremediated toxic site
Wastewater treatment plant

SAN FRANCISCO BAY: SEVEN-LAYER DIP

> The world shaped by capitalist modernization is not good for human life and is certainly rough on the health of the planet.
>
> —Rebecca Solnit,
> *Infinite City: A San Francisco Atlas*[1]

The San Francisco Bay shoreline has been a dump for as long as people have been living near it. For at least three thousand years, the Ohlone, Miwok, and Muwekma people located their middens near the water's edge. Later, during the Gold Rush of the 1840s and '50s, miners used mercury to extract precious metal flecks from the Sacramento and San Joaquin riverbeds, and that mercury—along with millions of tons of dirt and gravel—washed down the rivers and into the bay's mud, where it remains. Well into the 1900s, bayside boomtowns killed two birds with one stone by throwing their household trash and construction debris, and whatever else they could find, into the "wasted" space of mudflats and low marshes, using their cast-offs to create more land on which to build industrial facilities, airports, landfills, roads, offices, and neighborhoods. By the start of the 1960s, one-third of San Francisco Bay, the West Coast's largest estuary, had been filled.[2]

Around the same time that people stopped throwing their solid waste into the bay, electronics manufacturers started pouring their liquid waste in. For several decades in the twentieth century, Silicon Valley companies discharged industrial wastewater full of PCBs, heavy metals, and hundreds of other toxins directly into the shallow bay. Clean air and water legislation passed in the 1970s put a stop to the most egregious acts, but illegal dumping and unlawful discharges never fully went away.

Fast-forward a few generations to 2016, when Baykeeper, an environmental watchdog that tracks pollution in the Bay Area, filed a class-action lawsuit against the local landfill, Newby Island Recovery Park, over its long-term, consistent release of pollutants, including selenium, iron, aluminum, nitrogen, and sediment, into local creeks. Located in the once-tidal marshlands of the South Bay, Newby Island is the last

Younger generations report more climate change concern and action than older generations.

Global warming is personally important

Millennials	Gen X	Boomers	Silent
73%	69%	63%	58%

Willing to contact government officials about global warming

Millennials	Gen X	Boomers	Silent
37%	32%	30%	28%

Willing to donate money to an organization working on global warming

Millennials	Gen X	Boomers	Silent
38%	35%	30%	30%

Willing to volunteer for an organization working on global warming

Millennials	Gen X	Boomers	Silent
38%	33%	28%	20%

by wealthier and whiter students. Once in the workplace, the Deltas irritated their bosses with their "you broke it" attitude and endeared themselves to those same bosses with their "we'll fix it" ethos. For public health safety and for their carbon budgets, they stayed closer to home than previous generations had: telecommuting as much as possible, flying less, and staying longer when they did travel. Always a little leery of the latest whiz-bang products and technologies, they spent a lot of time enjoying the Bay Area's nature that previous generations had fought to protect and restore. From dump to divine, the Alphas are living their best lives around the bay.

Part II

STRENGTHENING STORMS

There is a fundamental idea in psychology and medicine that the path your life takes depends on just three causes: how you manage your challenges, protect your vulnerabilities, and increase your resources. These causes are located in three places: your world, your body, and your mind.

—Rick Hanson, *Resilience*[1]

This is how storm trauma unfolds. Meteorologists detect an atmospheric disturbance and attempt to predict its strength, speed, and direction. A storm is coming. Local officials begin broadcasting the urgent message, "Prepare to evacuate. You risk your life by staying. You risk the lives of disaster service workers by staying. Get out now." Most people flee, some freeze, some stay and vow to fight. Twenty-four-hour news media allow everybody to watch the lashing winds, the booming waves, the rain everywhere. We wait and hope. Once the storm has passed, it leaves in its wake ravaged spaces where houses once stood; dark nights with no electricity, heat, or running water; isolation as phones fail and exit routes flood. The first bodies are discovered, and everyone knows that more are out there. Everything is wet, and the first spots and smells of mold appear. Frontline workers begin the process of triage and assessment, as a new reality sinks in—a reality that can continue sinking in for months or years to come.

Whereas sea-level rise has been called the "slow emergency" of climate change, storms are the "fast" emergencies, the sudden disasters and attendant traumas that wreak unanticipated havoc on human and ecological systems. Storms are violent disturbances in the atmosphere that have just three ingredients—moisture, energy, and wind—and climate researchers focus on three basic storm types: tropical cyclones, extratropical cyclones, and thunderstorms.

Tropical cyclones—also referred to as hurricanes or typhoons—are born at sea. They are picky, requiring a very specific combination of ingredients: warm ocean water, low wind that blows at just the right speed and altitude to lift up that warm water, and another layer of light winds higher in the atmosphere that blow in a circular formation. Tropical cyclones build strength over the open ocean but weaken when they encounter land. This is why they are especially dangerous for islands, which are often surrounded by all of the ingredients of great storms but aren't large enough to dissipate their force.

In the United States and Europe, hurricanes are categorized by their maximum sustained wind speed. A Category 5 hurricane clocks wind speeds over 157 miles per hour, a Category 4 has speeds between 130 mph and 156 mph, and so forth. Climate scientists have recently speculated that a new category—Category 6—may need to be added due to the increasing intensity of storms.[2] Sometimes wind speed is the critical factor in the damage that a hurricane inflicts, as was the case with Hurricane Maria in Puerto Rico, where ferocious winds took down the electrical grid. Sometimes it's the sheer amount of rain that causes the problems, as with the five feet of rain that fell on Houston, Texas, over four days during Hurricane Harvey.

Extratropical cyclones begin in the middle latitudes and can form over both land and sea. Blizzards, nor'easters, and run-of-the-mill rainstorms all belong to this category. Such events can strengthen or dissipate rapidly depending on local geography and weather conditions, and they drive much of the earth's day-to-day bad weather. Most storms in northern Europe and along the East Coast of North America fall under this umbrella, and many large port cities are vulnerable to both river flooding and oceanic storm surge, often at the same time.

The National Oceanic and Atmospheric Administration (NOAA) has observed that all types of storms have become more intense over the past few decades, and predicts that storms will continue to increase in intensity as a result of climate change.[3] Because warmer air carries more water, clouds can hold more precipitation, which can cause more flooding. While storm *intensity* (heavier rain, faster wind, longer duration, and bigger ocean swells) is already increasing, the picture for storm *frequency* is less clear. Some climate models predict that tropical cyclones will not become significantly more frequent but will become more intense, while other models predict both greater intensity and an increase in frequency.[4]

Regardless of which model proves more accurate, increased intensity leads to greater loss of life, property damage, and destruction of crops and livelihoods. Being able to predict the intensity, frequency, duration, and trajectory of storms has become a matter of urgency. Coastal areas in particular are under close scientific scrutiny, since 40 percent of the world's population lives within 100 kilometers (62 miles) of the ocean, and two-thirds of the world's megacities are located at

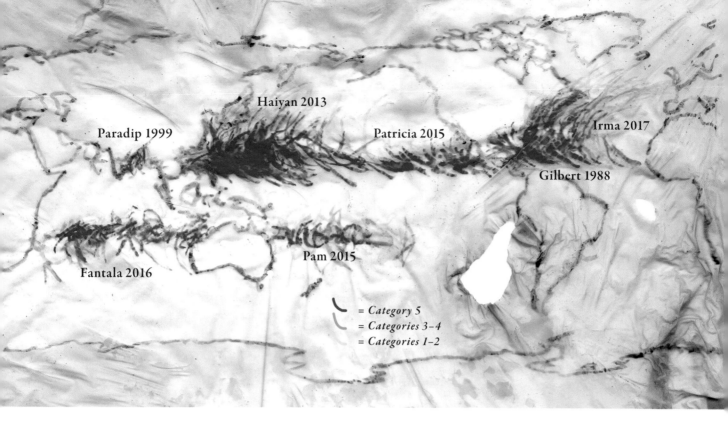

Haiyan 2013

Paradip 1999

Patricia 2015

Irma 2017

Gilbert 1988

Pam 2015

Fantala 2016

= Category 5
= Categories 3–4
= Categories 1–2

150 years of cyclone paths, plus strongest named storms since 1979

the coast. The percentage of coastal dwellers is expected to increase in the coming decades as humanity continues to urbanize.[5]

In a world of strengthening storms and rising seas, how and where we build cities is one of the most important questions we face this century because it will determine who is protected and who is in harm's way. Historically, poor people and immigrants have been located in the flood-prone, contaminated, poorly resourced urban areas. We have the opportunity to change that through urban design. Planners and public health officials have tools these days to understand current and potential risk by first identifying hazard risks (flooding, toxic spill, cliff failure), then extrapolating potential traumatic outcomes (drowning, acute asthma outbreaks, loss of homes), and finally assessing which factors

make a given population more likely to experience those outcomes (socioeconomic status, educational attainment, age, linguistic isolation, access to a vehicle, housing characteristics, and so forth).[6] Based on this kind of "vulnerability assessment," urban planners develop policies and plans to limit individuals' and communities' *risk* of experiencing various anticipated traumas, and to increase their *resilience*, their ability to bounce back, should trauma occur.

Deep inequalities exist between different communities' abilities to plan for storm response and recovery. Wealthy cities have stable populations that are easily tracked and counted. They also have the money and expertise to put in place protective measures like early warning systems; infrastructure like tide gates; long-term policies like zoning regulations with enforcement; and community awareness programs to educate residents and officials. All of these actions can make a place more resilient to storm trauma. Developing regions have far fewer resources for community research, regional planning, and sea defenses, which increases the vulnerability of their populations.

One universal challenge is that political borders do not always overlap tidily with community or ecosystem boundaries. School districts, police precincts, road and transit authorities, city planning agencies, and state and national regulators each have their own sets of programs, funding, and staff, which have been allocated to serve the needs of a geographically bounded group of constituents. But disasters strike across these lines, and our response requires pooling resources and coordinating deeply between bureaucracies that were not set up to be pooled or coordinated and often have explicit rules against working together. Addressing climate change disasters will require sustained cross-boundary collaboration over years, decades, and generations. How we understand our vulnerabilities, losses, and opportunities can help frame our ability to adapt.

Traumatized Bodies

A well-functioning city responds to a hurricane by marshalling all its resources to the sites of greatest need. Similarly, when our bodies are injured, they respond with a flurry of activity. Our blood vessels expand and become more permeable, which allows the site of the wound to be flooded with "first responder" cells like antibodies, phagocytes, and clotting agents. Sudden inflammation follows, which is essential to repairing damaged tissues, but it can also damage surrounding areas, a result doctors refer to as the "innocent bystander effect." Large wounds and major infections can lead to "cytokine storms" in the immune system, in which runaway swelling overwhelms our bodies and leads to life-threatening complications. Even after the initial crisis is over, organs can fail and infections can fester. Total system breakdown is a real possibility.

These same kinds of impacts occur in the aftermath of a storm. Emergency responders flood to the site of greatest need, even as community-wide communication systems may break down. The larger the storm and the less resilient the society, the greater the chance of system failure. However, those with stronger infrastructure, transportation, and institutions (read: bones, nerves, and organs) can restore healthy function more quickly. Such storms also rip through habitats below the waterline, harming animals and ecosystems. Fish usually swim away from a storm's path, but slow-moving invertebrates and corals are especially prone to devastation.

When hurricanes or cyclones repeatedly pummel a region, it has an increasingly difficult time recovering full function. In the same sense, repeated or extended stress in our bodies is associated with many diseases, including heart failure, kidney disease, type 2 diabetes, obesity, Parkinson's disease, depression, and anxiety. Chronic stress causes our body's emergency response system, called the sympathetic nervous system, to stay turned on for long periods, priming us with hormones like cortisol and andrenaline that cause cellular damage.[7] Biologists have shown that prolonged stress even impacts how our genes are expressed (or not), changing our bodies over generations—just as a city is forever marked by its past policies. As more frequent floods occur in geographically vulnerable places, coastal cities will face depopulation, lost revenue, and the vast expense of rebuilding. Decisions about what to keep, rebuild, or abandon will abound, sometimes made more difficult by our past and present planning choices.

Many of us don't like thinking about trauma in advance, because it promises loss, pain, and disorientation. But opening ourselves to the nature of impermanence and vulnerability can be a real asset as we consider how to live in and through the climate crisis, which will be punctuated by flood, fire, and famine. As both individuals and societies, how we prepare for trauma emotionally, socially, and financially will have everything to do with how successful we are. Leaders can strengthen community foundations by investing in resilient infrastructure, facilitating civic awareness and engagement, and retooling how bureaucracies work so that they are more nimble in times of crisis. Such preparations will not only save us pain; they will also save us money: one recent federally commissioned study in the United States found that for every dollar spent now on hazard mitigation, taxpayers save $6 in future disaster-relief costs.[8]

Recovering from injury doesn't mean going back to the way things were, because trauma and healing inevitably change the trajectory of our bodies, minds, and spirits. Trauma specialists study post-traumatic stress disorder—when the past haunts the present and future—as well as post-traumatic growth—when difficult events propel a person toward greater insight and strength for the future. All people and places have the potential for both.

HOUSTON:
WE HAVE
A PROBLEM

Sometimes you have to look at Mother Nature and say
"Mother Nature wins."

—Ed Emmett, County Judge of Harris County, Texas[1]

Nighttime shots from the International Space Station make Houston look a little like a roulette wheel. In the center of the wheel spins the urban core, bright parkways and thoroughfares radiating into the dusky suburbs. This aerial view sparkles and captivates, and the lines and curves of lights create a pleasing, orderly pattern on the Galveston bottomlands. Daytime views of the same scene offer a different story. The Houston Ship Channel, gateway to one of the busiest seaports in the United States, snakes its way through a messy landscape of factories, refineries, heavy industry, and, incongruously, homes. More than two hundred chemical facilities and a quarter of the nation's energy-refining capacity are situated along the Houston Ship Channel.[2] One of the largest petrochemical complexes in the world, these facilities are interspersed with low-income and minority communities whose health is impacted daily by their industrial neighbors and whose lives are particularly at risk during storms.[3]

In 2017, Hurricane Harvey was a stark reminder of the risks to society, environment, and economy that industrial cities face from climate change. The hurricane also happened to be the third "five-hundred-year storm event" in three years (such a storm event has a one in five hundred probability of happening each year). But just two years later, another five-hundred-year storm—Tropical Depression Imelda—hit the same region and caused an estimated $5 billion in damage.[4] In what was the wettest tropical storm ever to hit the United States, Harvey dumped five feet of rain on Houston over four days in August

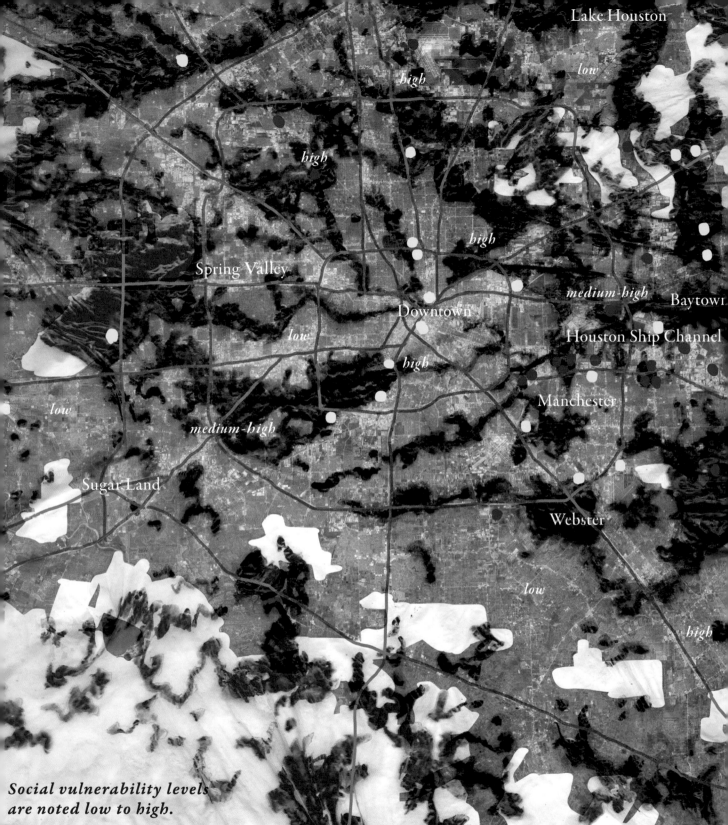

Lake Houston

high   *low*

high

high

Spring Valley

medium-high   Baytown

Downtown   Houston Ship Channel

low

high

medium-high   Manchester

low

Sugar Land

Webster

low

high

**Social vulnerability levels
are noted low to high.**

○ Superfund sites

● Petrochemical plants

● Greater Houston urban area

● Hurricane Harvey flood zone

○ Rural Texas

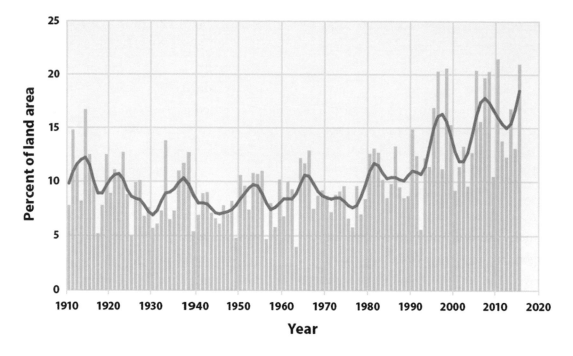

Each bar represents the highest one-day precipitation event for each year from 1910 to 2015. The line is a nine-year rolling average.

2017. Climate models project that these intense events will increase in the decades to come.

Houston's position in the middle of Hurricane Alley had made researchers and planners uneasy for years before Harvey struck. Studies conducted by the National Oceanic and Atmospheric Administration (NOAA), Rice University, Texas A&M, and the University of Texas had all issued warnings about the perils of Houston's petrochemical industry.[5] Author Roy Scranton had published an article in the *New York Times* a year before Harvey with a spookily prescient description of what a major hurricane would do: "If a storm rips through the region, it will hit an area that includes roughly one third of the country's known oil and natural gas reserves. . . . The refineries and plants encircling Galveston Bay are responsible for roughly 25 percent of the United States' petroleum refining, more than 44 percent of its ethylene production, 40 percent of its specialty chemical feedstock and more than half of its jet fuel."[6]

Local residents took to social media during and after Harvey to describe "unbearable" chemical smells from Exxon's refinery, but the company only disclosed its pollution releases two days after the storm had passed. Failure to comply with safety standards is practically routine for refineries operating in the region, and plant fires happen with distressing frequency. The watchdog Environmental Integrity Project concluded from public records that the storm had caused petrochemical plants to release at least 8.3 million pounds of unpermitted air pollution.

Residents reported flares, leaks, chemical discharges, noxious odors, and fires at the Arkema chemical plant in Crosby, just outside of Houston.[7]

One way to check environmental risk is to price real estate: the cheapest real estate is usually in the riskiest areas because those who have the ability to choose will not live on top of a toxic dump or in an area that floods and festers. This is one route by which environmental injustice becomes intergenerational: poor people can't afford to move to somewhere less risky. In Manchester, a neighborhood in Houston's East End, nearly one-third of residents live below the poverty line; 89 percent of residents are Hispanic, another 6 percent African American.[8] Manchester is surrounded by Port Houston, two abandoned Superfund sites, and ten rubber, gas, and chemical facilities that collectively pump out 1.9 million pounds of air pollution every year.[9] More than half the Superfund sites flooded by Harvey are in disadvantaged neighborhoods.[10] Local residents are paying with their lives so that the rest of the country can pursue an American Dream of chemical-intensive consumption.

Those who have the fewest resources to cope with the traumatic impacts of natural disasters are often the ones who are most frequently exposed. People who don't own a vehicle, because of disability, finances, or other life circumstances, are less able to evacuate, and those without family or friends who have money or space to spare might have nowhere to go even if they could get out. During a disaster, recent immigrants are often less likely to seek help from government institutions and first responders than are long-term residents.[11] People who wind up in emergency shelters—with all of the attendant difficulties of overcrowding and limited access to medical services—are disproportionately low-income and people of color.[12]

Recovery from natural disasters can also be more difficult for renters and people living in public housing who are reliant upon landlords and government bureaucracies for remediation of problems like mold, structural damage, and exposure to hazardous materials. Remediation in public housing can take years, if it happens at all. And, as a final injustice, Rice University researchers found in 2018 that natural disasters actually increase the wealth gap in the United States, concluding, "The broader implication is that two major social problems of our day—wealth inequality and rising costs of natural hazards—are connected in ways that involve not just how we develop and administer policies related to incomes and finance but also environmental hazards."[13]

In the United States, most land-use decisions are made by local governments. This means that communities can create spaces that reflect the values and the preferences of those involved in the local planning process. Historically, the locals in Houston who made the plans, passed the laws, built the spaces, and opened the facilities were white. As Professors Robert Bullard and Beverly Wright point out, the people who have been subjected to the worst impacts of those local land-use decisions are African Americans and waves of immigrants from Asia and Latin America—all of whom, they say, have "the wrong complexion for protection."[14] Most public housing was built in flood-prone areas, where natural drainage is fragmented and some neighborhoods lack both sidewalks and proper infrastructure to funnel water off the streets. Residents of those neighborhoods were exposed to the contaminated soup of raw sewage, industrial discharge, and other chemicals after Harvey's rain left water pooling on the land for days and weeks. Poor communities have historically been at greatest risk because they have been less likely to hold the political power to impact land use and environmental health decisions. A new generation of environmental justice leaders is emerging and giving communities a stronger voice in order to end the marginalization and exposure to environmental hazards that previous generations have struggled against.[15]

～～～～～～～～～

Key Term: Risk

The term *risk* can carry a note of excitement for those who enjoy the feeling of an adrenaline rush, and the origin stories of the United States frequently employ a risk-and-reward narrative: the "Manifest Destiny" of the pioneers on the plains, the '49ers in California, the hellcats in Texas. Those narratives ignore the racism and rapaciousness of those episodes; the sacrifices and losses are buried. History has been written by the victors.

More accurately, both as a noun and as a verb, *risk* relates to the probability of an event's happening and the potential loss

connected to it. Risk management is the practice of identifying, quantifying, and then limiting threats to a person, program, system, or organization. Because thorough risk management requires an armada of data, most individuals don't (or can't) assess their own risks realistically. So while land-use decisions may be local, risk management decisions are not. These decisions about whom to insure, what to prioritize, where to focus, and how to proceed despite uncertainty are the purview of government agencies, working with for-profit financial institutions. Risks are packaged, privatized, and then publicly sold in the form of bonds. Government bonds are well-known vehicles for managing risk, and newer forms of "cat bonds"—catastrophe bonds—have created a market betting on all kinds of climate change risks, from collapsing fish stocks to major flooding to government collapse to global pandemics.

～～～～～～～～～

A View from 2050

As Kai watched the team of MIT researchers load their gauges, nozzles, and canisters marked "CAUTION—BIOHAZARD" onto his barge, he wondered just how their lab-born idea would work here in the real world of post–Hurricane Ada Houston. He'd heard of MIT's new miracle technology, of course; all port employees working in disaster response had been required to attend a safety seminar last month, during which the head of the Storm Remediation Technologies Lab described "The Fix." The research

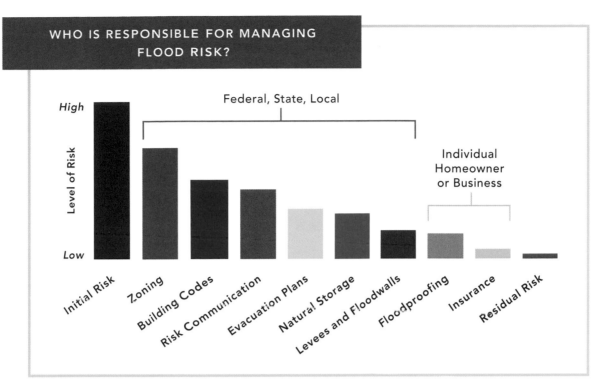

WHO IS RESPONSIBLE FOR MANAGING FLOOD RISK?

Federal, State, Local

Individual Homeowner or Business

Level of Risk — High / Low

Initial Risk · Zoning · Building Codes · Risk Communication · Evacuation Plans · Natural Storage · Levees and Floodwalls · Floodproofing · Insurance · Residual Risk

Flood risk is managed by a complex set of local, state, and federal government rules and programs, along with public and private insurance and individual actions.

team wanted to take The Fix to market, and Houston was the perfect product testing ground. But the lab director had wandered into a forest of technical jargon before most of the attendees had grasped just exactly how this high-tech Fix was going to make their pollution problems go away.

Today was finally showtime, and Kai was excited to find out. One of the lab techs came over, and they checked the map on her GPS unit against Kai's barge instruments. The communications director herded all of the VIPs who had been invited to the unveiling on deck: the mayor, a congressperson, donors to MIT's lab, national and local press. As the barge slowly chugged up the Ship Channel, the techs misted the

contents of the canisters over the water, spraying from Barbour's Cut in the south to Boggy Bayou mid-channel. The Fix, it turned out, was bugs. In a petri dish in a lab somewhere along MIT's Infinite Corridor, a team of post-doctoral researchers had discovered a microbe with an affinity for hydrocarbon. "These microbes—tiny bugs—they *eat* the oil," explained the communications director, trying to summarize in lay terms what it had taken the team of microbiologists the better part of a decade to discover. At the pre-selected coordinates, Kai reversed the barge, the techs unhooked the sprayers, recoiled the tubing, and recapped the canisters. The photographers shot a few last half-hearted photos, the VIPs

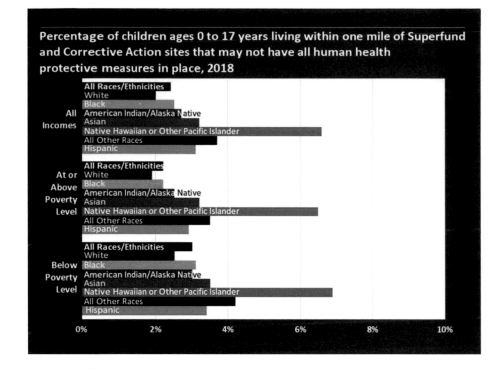

White children are the least likely to live in proximity to federally recognized contaminated sites.

Percentage of children ages 0 to 17 years living within one mile of Superfund and Corrective Action sites that may not have all human health protective measures in place, 2018

All Incomes
- All Races/Ethnicities
- White
- Black
- American Indian/Alaska Native
- Asian
- Native Hawaiian or Other Pacific Islander
- All Other Races
- Hispanic

At or Above Poverty Level
- All Races/Ethnicities
- White
- Black
- American Indian/Alaska Native
- Asian
- Native Hawaiian or Other Pacific Islander
- All Other Races
- Hispanic

Below Poverty Level
- All Races/Ethnicities
- White
- Black
- American Indian/Alaska Native
- Asian
- Native Hawaiian or Other Pacific Islander
- All Other Races
- Hispanic

0%   2%   4%   6%   8%   10%

shook hands with each other, and everybody disembarked. They would have to wait a week to see if The Fix worked.

The bugs were supposed to digest the slicks of chemicals floating on the water that had leaked from two refineries during last month's Hurricane Ada. Kai recalled the communications director explaining that the bugs would land on the slick, ingest molecules of oil, and then sink to the bottom because their new chemical composition would make them heavier than water. At that point, explained the comms director, it would be a simple matter of scooping up the heavy bugs and the mud using the same equipment that was already deployed to dredge the channel for shipping.

The next morning, Kai's display showed that, in fact, the bugs were already settling on the mud floor of the channel. The microbes showed up on his sensors as specks of orange, a dot-picture of a snake slithering up the Ship Channel. Fewer than half of the bugs were still suspended in the water column of the river, and the MIT team assured the Port Authority that the majority of the remainder would continue to settle downward as the churn from the recent hurricane subsided.

Although dredging practices had come a fair way over the past forty years, and newer protocols incorporated plenty of silt scrubbers, recapture filters, and microbrooms, oil-spill remediation was still a messy business. Dredging always stirred the pot and released potentially lethal chemicals back into the water column, from which it could get to all kinds of places. Still, Kai reflected, if mud scrubbers and giant cans of microbugs could prevent even half of the birth defects and cases of leukemia, skin rashes, and neurological problems that the local kids had suffered for so long, he'd call it a win.

Kai had grown up in Houston's East End after Harvey, during the terrible years when his grade-school friends would show up with thick red rashes and persistent wet coughs. Some had left public school to be home-schooled to limit their exposure and structure their learning around their chemo appointments. Kai had attended funerals of two classmates while he was still in high school. The polluters in the region had long used the same playbook as the tobacco companies had in the prior century. Deny. Profess ignorance. Fund studies conducted by your own staff. Deny some more.

By the early 2030s, the persistent chemical leaks and spills from semi-annual floods had been investigated by the EPA, and the jig was up. The debate was over, the polluters needed to pay. Some of the biggest petrochemical refining companies had even fast-tracked the culpability-payment-bankruptcy-restructuring-profitability cycle, and were back in operation by 2035 with shiny new names and corporate mission statements that included words like *sustainability* and *community care*. None of that window-dressing brought back the people, the animals, or the neighborhoods.

But, Kai reflected, beggars couldn't be choosers. Hurricanes came faster and harder now, and the city wasn't keeping up. Although many people had left the Houston area, and plenty of industries had attempted to hurricane-proof the infrastructure that ringed the gulf, change was slow and relocation was hard. City leaders, first responders, and the industrial facilities had felt confident heading into Hurricane Ada that everything possible had been done to batten down the hatches. Many of the residents weren't so confident, however, and their repeated and very public attempts to get the city to do something more had caught the attention of the MIT researchers, who were looking for just such a case to test The Fix. As Kai watched the orange dots drift slowly across his display, he wondered what the collateral damage from Ada and The Fix would be.

Hamburg sits at the edge of a broad, glacial
valley carved at the end of the last ice age,
around ten thousand years ago.

North Hamburg

Central Hamburg

Elbe River

HafenCity

North Elbe

Port of Hamburg

Wilhelmsburg

South Elbe

Harburg

- Urban area
- Rural area
- Land projected to be below
  10-year flood level in 2060
- 1 meter flood level in 2020
- 3 meter flood level in 2020

HAMBURG, GERMANY: RIVER CITY AT RISK

Investigate what is, and not what pleases.
—Johann Wolfgang von Goethe,
The Attempt as Mediator of Object and Subject

The Elbe River starts in the Krkonoše Mountains in the Czech Republic, wends its way across Germany, and finally empties into the Wattenmeer (Wadden Sea). When North Sea storms roll in, tidewaters make the reverse route back up the river seventy miles and into the Free and Hanseatic City of Hamburg. This North German city has endured severe flooding from the North Sea since Karl der Grosse (Charlemagne) built a castle here in 808 C.E. and will continue to flood as long as Hamburgers eat wurst. It's this region's wealth of resources—economic, social, environmental—that has so far allowed the city to recover its former functioning after past storms and plan for resiliency to the storms of the future.

Like their Dutch neighbors, Germans have spent the past several centuries trying to engineer their way out of disasters brought about by the temperamental conditions of the North Sea region.[1] Since at least the 1750s, Hamburg has invested in traditional flood barriers, including seawalls, dikes, and floodgates. However, these systems failed on the night of February 16, 1962, when a storm pushed 5 meters (18 feet) of North Sea water up the Elbe and over the dikes, submerging one-fifth of the city, destroying sixty thousand homes, causing the present-day equivalent of €1.6 billion in damage, and killing 315 people. That flood, the most expensive in the city's twelve-hundred-year history, galvanized local, state, and federal governments and spurred a total overhaul of coastal defenses that cost more than the equivalent of €2.2 billion over several decades.[2] Along with traditional flood barriers, Hamburg also began investigating creative approaches to protecting its citizens by "living with water."

NEW YORK, NEW YORK: CAPITAL OF CAPITAL

The wetlands are migrating. Can we welcome them?
—Ayasha Guerin, New York artist[1]

Like people, cities demonstrate their essential character when responding to a crisis. As a collection of islands and one-time swamps where the Hudson and East Rivers meet the Atlantic Ocean, New York was reminded of its basic geography in 2012, when Hurricane Sandy's catastrophic surge brought an extra three meters (ten feet) of seawater into the estuary. Initially, Sandy was thought to be a rare event that occurs only every seven hundred to one thousand years. But researchers at Columbia University found that the return period for a Sandy-size storm is just 103. With the one meter of sea-level rise predicted by later this century, such inundations may occur as often as every twenty-eight years, with smaller but substantial floods interspersed.

Coastal megacities are in for an especially difficult journey as seas rise and storms become more intense.[2]

In his major post-storm speech, Mayor Michael Bloomberg noted that in 2050 one-quarter of the city's land and 800,000 residents would be within the one-hundred-year flood zone. But instead of talking about the devastation as an opportunity to reshape the city's shoreline to better reflect future sea levels and more frequent storm surges, he doubled down. "As New Yorkers, we cannot and will not abandon our waterfront. It's one of our greatest assets. We must protect it, not retreat from it," he said.[3] He pledged $20 billion in new fortifications, promoted huge luxury developments in the flood plain like Hudson Yards, and suggested creating

Social vulnerability

Most vulnerable

Least vulnerable

New Jersey

New York

The Bronx

Newark Airport

Jersey City

Manhattan

LaGuardia Airport

Long Island

Queens

New Jersey

Brooklyn

JFK Airport

Staten Island

Verazzano Narrows

Edgemere

Painted areas show
5 meter flood line

Middletown, New Jersey

more infilled land beyond the city's current shoreline.

New York's state government took a different post-Sandy approach, buying out some homeowners in Staten Island, demolishing their houses, and not permitting any future development. A federally funded design competition, Rebuild by Design, generated half a dozen innovative projects to improve rather than merely replace what had been there before, such as oyster beds to absorb wave intensity off southern Staten Island. One of the projects, a massive sea wall around the tip of lower Manhattan, is currently under construction and will, in fact, make these neighborhoods safer and more desirable—which will then encourage more growth and higher rents. It will also refract future storm surges onto the shorelines of Brooklyn, Jersey City, and Hoboken, which are already more vulnerable than Manhattan in both topography and demography.

But funding and bureaucracy pared down and slowed their construction. "Everyone was working at cross purposes," according to Liz Koslov, an expert on post-Sandy recovery at the University of California, Los Angeles.[4]

Bloomberg's successor, Bill de Blasio, took long-range planning a step further by creating a Resiliency Office to coordinate upgrades to the city's infrastructure and zoning rules and to oversee the reduction of city emissions by 80 percent, from 2005 levels, by the year 2050.[5] However, he shares his predecessor's basic assumption that New York's footprint is simply too important to question or change. So the city continues to reinforce its historic economic and development priorities, relying on modest changes to land-use policies and building codes to keep people safe. People are not demanding change, so "there is no real desire to reverse course," says Koslov.[6]

Some recent adaptation plans have taken a more comprehensive approach, but even these efforts aren't solving for the long term. The Resilient Edgemere plan will spend $100 million to improve a small, neglected community on Jamaica Bay in Queens with new low-income housing and infrastructure, conversion of a few flood-prone residential blocks into a park, a voluntary home buyout program, and an earthen berm to protect against thirty inches of sea-level rise. Yet the plan acknowledges that average sea levels will be higher than this in just a few decades—not including any unusually large storms or tides. So, expensive half measures are taking the place of difficult conversations and unpopular zoning changes. "No one is building for the 500-year storm," says Lynn Englum of Rebuild for Design.[7]

New York does have its share of truth tellers who are trying to get the attention of politicians and the public. "Sandy may not happen again like it did," according to J. Andrew Martin, acting chief of the Federal Emergency Management Agency's (FEMA) risk-analysis branch in New York, "but there will be something very similar, and it's not that far off."[8] Klaus Jacob, of the Earth Institute at Columbia University, has warned, "Sea-level rise is the ultimate fate of the city. . . . It *can* adapt. [New York] can be a totally lively city

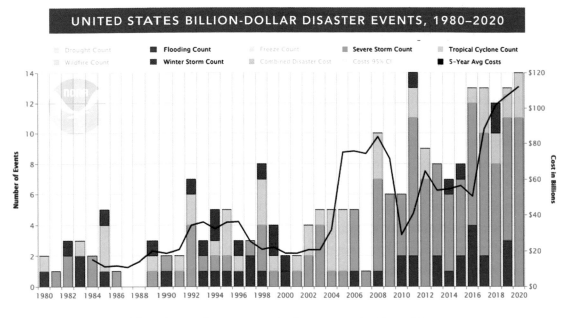

UNITED STATES BILLION-DOLLAR DISASTER EVENTS, 1980–2020

Storms and floods are the most costly disasters in the United States, costing on average $100 billion a year over recent years.

without [storm] barriers. We are lucky. We have high ground within our boundaries. We just have to move together on higher ground."[9] A city that has grown weary of disasters could begin investing in the future now.

At the federal level, the Army Corps of Engineers, which holds national responsibility for building and maintaining flood defenses, is studying several options to protect the greater metropolitan area, including a $119 billion moveable storm barrier reaching across the Verrazano Narrows. According to environmental watchdogs, this would be a catastrophe for the region's water quality and aquatic life, while the more modest and affordable options would provide less protection.[10] But none of these visions address the unyielding reality of

sea-level rise, which is predicted to rise one to two meters (three to six feet) this century. Though "strong local voices" make a big difference in the Corps' ultimate decision-making process, according to Brigadier General Peter Helmlinger, there is still little public debate on how to prepare for the inevitable.[11]

Another major federal player is the National Flood Insurance Program (NFIP), which sells mandatory insurance to property owners in the one-hundred-year flood zone. Its rates are only half what the market rate would be for similar insurance, which encourages overdevelopment in risky areas. And its benefits flow disproportionately to upper-income people and homes that flood repeatedly; 3 percent of the claims soak up 30 percent of the budget, and the program

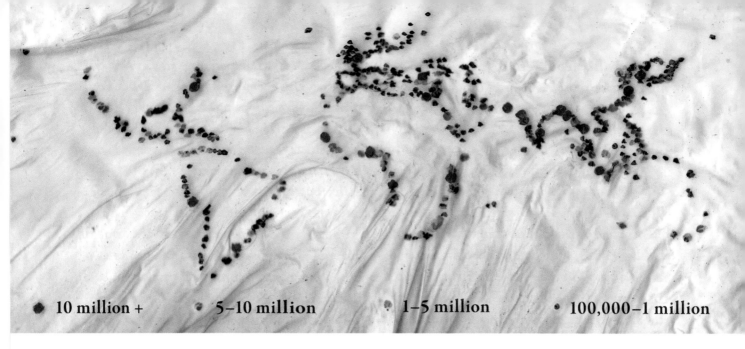

10 million +   5–10 million   1–5 million   100,000–1 million

Urban population centers at risk of flooding in the 2050s

But some communities on Staten Island, which had a lot of experience with flooding, had been prepared. They had held community meetings since Sandy, where they discussed and debated different approaches to coping with the Big One they all knew was coming. When it did, they felt clear-eyed, informed, and self-empowered. Their block-by-block safety teams helped the elderly and disabled evacuate to agreed safe houses, and the rest headed for the well-appointed shelter they had campaigned for a decade earlier, built on the high ground of the former Fresh Kills landfill. The borough had the lowest death toll in the city, even though it was largely a former swamp.

New Jersey and Connecticut were also devastated, with many smaller rivers breaching their banks and washing away small towns. The three governors crossed political lines to finally implement a plan that had first been proposed back in 2017, creating a Regional Coastal Commission and vesting it with the regulatory powers to relocate low-lying communities and add protection to higher, densely populated places.[17] After two years of thoughtful development, the commission released the New Mannahatta Plan, a vision that more closely realigned the metropolitan landscape with its original seventeenth-century topography. In Manhattan, Canal Street was reverted to an actual canal, and Water Street became the shoreline and a floodable park. In Staten Island and Queens, whole neighborhoods were rebuilt further inland, the buildings torn down and converted into storm berms along the +2-meter shoreline. All along the tri-state coast, the RCC established floating communities, added ferries and water taxis, converted the most undefendable sites to open space, and built even denser urban hubs on high ground.

They opened Meadowlands National Park as a restored wetlands, to great fanfare.[18] As the subway faltered and failed after repeated inudations, the city re-elevated its trains, running them up the dry spine of Manhattan from the Bowery to Broadway and Central Park. Transformational adaptation had finally reached the shores of the Hudson.[19]

It took more than two decades to fully implement, but as the city enters the second half of the century, it has squarely faced its challenging geography. Now, New York is indeed the "dramatically reshaped city" Zarrilli predicted decades ago, though its resilience is tinged with notes of sadness, healing, and regret.[20]

Puerto Rico

Virgin Islands

Anguilla

Netherlands Antilles

St. Croix

Antigua and
Barbuda

St. Kitts and Nevis

Montserrat

Guadeloupe

Dominica

Hurricane Maria passed over many small island nations and territories before reaching Puerto Rico. According to a 2020 study, cyclone intensity is increasing by 8 percent per decade around the world, and Caribbean countries are at especially high risk. Seventeen of the twenty costliest Caribbean hurricanes have taken place since the year 2000.

SAN JUAN, PUERTO RICO: PODER, DESPACITO

> Central to the experience of trauma is helplessness, isolation, and the loss of power and control.
>
> —Judith Herman, *Trauma and Recovery*[1]

Hurricane Maria was born on September 13, 2017, as an atmospheric disturbance off the coast of Cape Verde, turning into a tropical storm while she traversed the Atlantic Ocean. Maria had intensified into a Category 5 hurricane by the time she struck the island of Dominica on September 18. She damaged or completely destroyed over 90 percent of the buildings on Dominica before heading northwest toward Puerto Rico. She slammed into Yabucoa, a town on the eastern coast of the island, on September 20 at 6:15 a.m. with wind speeds of 155 miles per hour.

Over the next thirty hours, Puerto Rico was drenched in thirty inches of rain, torn apart by "tornado-like" winds, and pounded by twenty-foot waves.[2] The island's electrical grid was destroyed, leaving all 3.4 million inhabitants in the dark and most people without access to clean water. President Donald Trump declared a state of emergency in the U.S. island colony on September 21 and then headed out for a weekend of golf.

His perfunctory response foreshadowed much of the federal government's reaction to the catastrophe. The year 2017 had already been a busy disaster season, with hurricanes Irma and Harvey wreaking havoc in the Gulf of Mexico and the Caribbean. By the time Maria made landfall, responders were tired, resources were tied up, and the president was seemingly uninterested. Puerto Rico therefore suffered two disasters, one natural and one man-made.

In the weeks and months after the storm, relief efforts were a tangled mess of bureaucratic bungling, strained resources, and confusion. Hurricane Irma, which had struck Florida two weeks earlier, had already depleted many emergency supplies, including "blue roofs"—temporary waterproof tarps that are nailed down where real roofs used to be. During the critical first week after each disaster, the federal government deployed three times as many on-the-ground personnel for Hurricane Harvey in Texas and twice as many for Hurricane Irma in Florida than they did for Maria in Puerto Rico. More than half of the on-the-ground responders in Puerto Rico were still trainees without proper certification to perform the jobs that they were doing.[3] The National Guard was redeployed elsewhere before they had even finished distributing food, medicine, batteries, fans, and other relief goods. These donations sat rotting in a parking lot for months.[4]

As hours stretched into days and then weeks, the power remained out across much of the island. Lieutenant General Todd Semonite recounted, "All of a sudden, about the eighth day in, the federal government asked us to step up and be able to take on this mission of grid repair. But it is not something that we planned on doing in any type of disaster. We don't do grid repair, usually, normally, doctrinally."[5] While federal disaster response agencies including FEMA, the Army Corps, and the National Guard all tried to figure out what they were doing, the lack of electricity led to hundreds more deaths. Without electricity, hospitals weren't able to adequately treat hurricane victims or administer life-saving procedures like dialysis. Nine days after landfall, the Department of Defense reported that, in a survey of 90 percent of the island's sixty-nine hospitals, "one is fully operational, 55 are partially operational, five are closed, and the status of eight is as yet unknown."[6] Without electricity, clean water could not be pumped from the wells that serve a large part of the island, and communication was nearly impossible. According to FEMA, "95% of cell towers in Puerto Rico were out of service and outages continued in the ensuing months. As a result,

local, territorial, and federal agencies faced difficulties knowing what was needed and where in the immediate aftermath of the storm."[7]

Meanwhile, the bodies piled up. The death toll reported in December 2017 by the Puerto Rican Department of Public Safety put the number of victims at sixty-four, but that count only included confirmed deaths that were a direct result of the storm (like drowning) and was roundly criticized for not including victims of secondary effects (like sepsis). Soon media outlets, academic institutions, and nonprofit organizations responded by publishing their own competing death tolls. The *New York Times* estimated that 1,052 had died in the first forty-five days after Maria, and researchers at Harvard estimated 4,000. Once all of the bodies that were stored in FEMA refrigerators had been processed (after additional medical examiners were flown in to support the locals who worked at the one and only morgue on the island), the official death toll was revised to 2,975.[8]

How is it possible that American citizens were left in such conditions, and for so long, and no one could figure out how many Americans had died? Because, in both law and in practice, Puerto Ricans are only sort-of-Americans. They can serve in the military but can't vote in presidential elections. They can elect their own House representative (called a Resident Commissioner), but that representative can't vote on the House floor. Tax law treats Puerto Rico as a foreign entity, but maritime law treats it as part of the United States. The result of all of this legal confusion is that Puerto Ricans have very little political or financial power.

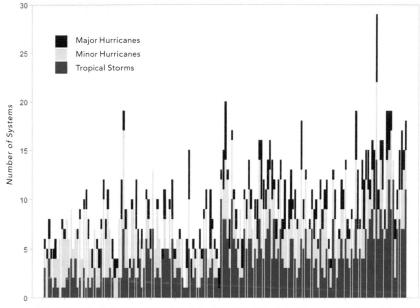

The number of tracked tropical storms in the North Atlantic basin from 1850 to 2014

This segregated status goes back to the Insular Cases, a series of early twentieth-century Supreme Court decisions that effectively cut off a path to statehood by declaring that "Puerto Rico belongs to but is not part of the United States" and designating the island as an "unincorporated territory." The Insular Cases are still on the books today, despite referring to the inhabitants of the subject territories as "alien races" and "savage" and stating that the cases uphold "principles of natural justice inherent in the Anglo-Saxon character."[9]

The island's inability to self-determine made Maria's impacts far worse. While federal government officials blame the island's geography for the slow response, either statehood or independence would likely have made emergency funding available more quickly. Contrast the island nation of Dominica—also destroyed by Maria—which entered into agreements with the World Bank, resulting in a $115 million package of support for recovery.[10] Without sovereignty, Puerto Rico has no such ability to negotiate, and all rebuilding and recovery are handled through the federal government in Washington, DC, fifteen hundred miles away. A year after the hurricane died out, Puerto Rico struggled to negotiate debris removal and perform autopsies on refrigerated stacks of corpses, while Dominica pursues a funded vision of becoming the first climate-resilient country in the world.[11]

In the aftermath of Hurricane Maria, Puerto Rican politicians have pushed harder than ever for statehood.[12] They argue that long-term recovery should include ending Puerto Rico's status as the "oldest, most populous colony in the world."[13] On-island Puerto Ricans have voted five times on the issue. In 2017, 97 percent of voters favored statehood, but voter turnout was low, as usual, and the matter didn't go anywhere. The Puerto Rican diaspora in the United States, which is mostly located in Eastern Seaboard cities, has tended to favor independence over statehood. Lack

Cox's Bazar District
Bangladesh

Myanmar

Bangladesh

Chin State
Myanmar

Kutupalong Camp
(630,000)

Other camps
(280,000)

Rakhine State
Myanmar

Naf
River

3 meter flood zone
Villages burned 2017–2018
Refugee camps (populations
Political borders

Bay of Bengal

In places like western Myanmar,
where strengthening storms,
floods, erosion, and crop failure
exacerbate existing ethnic persecution, increasingly the only
answer will be to flee.

Kaland
River

Sittwe

KUTUPALONG CAMP, BANGLADESH: HUMAN TIDES

The truth is, we are entering an age of migrants, and we must adjust our sense of fairness and morality, and even our concept of national borders, accordingly.

—Tahmima Anam, British-Bangladeshi novelist[1]

Setera Bibi was twenty-three when the soldiers attacked her village one night. They murdered her husband and burned her home. After walking for two days toward safety with her two daughters and her mother, she crossed the Naf River, which marks the border between Myanmar (formerly Burma) and Bangladesh, a daughter in each arm. The youngest, still a baby, was terrified, wrestled out of her mother's grip, and was swept away. When Setera's mother fell behind, soldiers beat her with a rifle and broke her back. Her mother was eventually able to reunite with her family in Kutupalong Camp on the Bangladesh side of the river, but she can walk no longer.

In just three weeks in 2017, half a million people like Setera hemorrhaged across the border from northwestern Myanmar into Bangladesh, overflowing camps that already housed tens of thousands of Rohingya who had fled past ethnic purges. Within a year, the total population of Kutupalong Camp reached 940,000, overwhelming aid agencies, and Setera's malnourished daughter became too weak to go to school because their ration of rice, pulses, and cooking oil simply wasn't enough.[2] The new camp residents had no way of growing food or making a living, and the newly deforested hillscape of bamboo huts was even more vulnerable to the torrential seasonal rains than their former floodplain villages. Sanitation was lacking, so disease and death patrolled the crowded paths, marked by bamboo sticks on the graves of

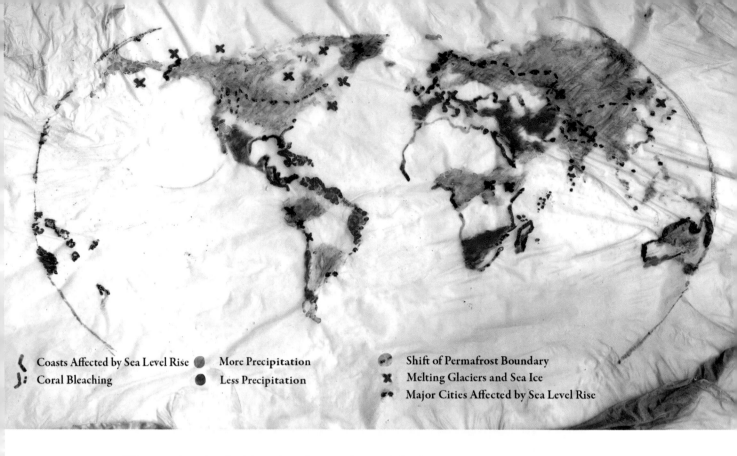

Coasts Affected by Sea Level Rise   More Precipitation   Shift of Permafrost Boundary
Coral Bleaching   Less Precipitation   Melting Glaciers and Sea Ice
     Major Cities Affected by Sea Level Rise

Climate migration is driven by changes in precipitation, temperature, and land and water use, as well as the coastal impacts of storms, erosion, and saltwater intrusion into farmland and drinking water.

geographic terms, and some disappear after a single storm; the United Nations has called Thengar Char "uninhabitable." Of the four million mostly poor Bangladeshis currently living on chars, most must relocate five to seven times in their lives due to erosion or flooding.[10]

Prime Minister Sheikh Hasina had said this new camp on Thengar Char (sometimes also called Hassan Char) would be a "temporary arrangement" to ease crowding in Kutupalong Camp. But her advisor, H.T. Imam, clarified that, once there, the Rohingya would only be able to leave if they were returning to Myanmar or were selected

for asylum by a third country. Imam said, "It's not a concentration camp, but there may be some restrictions. We are not giving them a Bangladeshi passport or ID card." Prime Minister Hasina further elaborated that, though the initial plan had been to put 100,000 people there, Thengar Char has room for as many as one million people.[11] Row upon row of close-set barracks, each housing sixteen families, have been built for a population that, once there, cannot leave, further magnifying their vulnerability. Bangladesh is not yet forcibly repatriating people to Myanmar, as they did in the 1980s and 1990s, but it has refused to accept any more refugees. In 2020, some young refugees have become so desperate that they are hiring human traffickers to take them to Malaysia, but both Malaysia and Bangladesh have refused the boats, leaving the refugees to starve at sea.

Nor can the Rohingya ever go home. Any returnees would be required to register as Bengalis (that is, foreigners), would not be granted return to their ancestral villages unless they could prove residency, would not be hired by distrustful Rakhine farm owners, and would fear for their safety every day. Myanmar appears by its actions to prefer homogeneity to international legitimacy, and Aung San Suu Kyi, the democratically elected leader, has taken no steps to assure the Rohingya's safety. So the repatriation centers sit empty.

In 2016, wars and violence displaced 9 million people within their own countries, while natural disasters displaced 24 million.[12] The legal distinctions between who is a war refugee and who is a climate migrant are becoming blurry as conflicts driven by climate disruption increase around the world. For the victims, layers of hardship and trauma simply accumulate, generation after generation.

〜〜〜〜〜〜〜〜〜〜

Key Terms: Refugee or Migrant?

According to the United Nations Human Rights Council (HRC), *refugees* are people fleeing violence and persecution in their home country by crossing into another. *Migrants*, on the other hand, are people who move to seek opportunities, either within or outside their home country. IDP status falls somewhere in between, with theoretically far fewer rights to aid or resettlement than refugees, but more rights and protections than migrants. Born in the aftermath of World War II, this hash of different statuses has created a hierarchy of suffering. Clearly current definitions are inadequate to twenty-first-century needs.

Many human rights advocates reject the term *climate refugee* because they know some countries would use this new definition of refugee to dilute their legal obligation to protect and assist people fleeing war and persecution.[13] The Environmental Justice Foundation distinguishes three types of climate migrants: those fleeing a disaster, those who are motivated to migrate because of climate upheaval, and those who are *forced* to do so by sea-level rise, persistent drought, and other climate impacts. The term *survival migration* is also used now because it acknowledges the mixed motivations that are central to these stories.[14]

One of the most important things the international community can do to ease suffering and slow the flows of people crossing borders is to come up with thoughtful and compassionate policies that reflect twenty-first-century realities. The Global Compact for Safe, Orderly, and Regular Migration (GCM), an international agreement that provides an important starting point, was debated and signed in 2018 by most countries in the world. Countries that have declined to sign include the United States, Italy, Iran, Australia, Chile, Israel, and several countries in eastern Europe.[15]

〜〜〜〜〜〜〜〜〜〜

A View from 2050

"Many refugee camps become permanent, urban spaces," says Fiona McCluney, a senior UN Habitat official, because they end up using the same layouts and roads that were slapped down in the midst of crisis. The UN and host country provide almost all the services and ongoing financial support, even though the host countries often have few resources themselves. Among the early orders of business in Kutupalong Camp was managing sewage, drinking water, and rainfall in the hilly landscape. So the United Nations' Development Program hired laborers to deepen some ravines into drainage canals both through and around the camp and to level much of the rest of the land to mitigate the worst monsoon flooding. Main roads were paved, with sewers beneath. Throughout the 2020s, the appearance of permanence grew as schools, hospitals, and community centers were improved from tarp and twine to breezeblock and brick.

By 2026, China had completed new road and rail links from Kunming to Chittagong, so Kutupalong Camp went from being the back of beyond to an important notch in the Chinese Belt and Road. It turned out that the camps were on higher, safer ground than many of the surrounding towns in Cox's Bazar District, so tens of thousands of Bangladeshis moved inland over the next two decades, as trade arrived and jobs followed. Meanwhile, in the Hague, where in 2019 the Gambia had asked the International Court of Justice to prosecute Myanmar for the Rohingyan genocide, the wheels of justice moved slowly.[16]

In the 2030s, a few countries that had once been homogenous, like Japan, Korea, Switzerland, and Finland, opened their own borders a crack, increasing their refugee quotas and offering financial assistance to far-flung regional powers that agreed to host climate migrants. This approach ultimately cost donor countries and international aid agencies less than it had to maintain traditional refugee camps, and brought needed revenue to the host countries. It also allowed people to stay closer to their home countries, families, and neighbors, making education, employment, and communication less burdensome for all. The UN Special Envoy on Migration worked with more wealthy countries to fund this decentralized strategy and better serve the tide of humanity flowing from the Global South. In the developing world, where 85 percent of global refugees lived, countries followed Uganda's early lead, allowing refugees to move freely, work, and contribute to the economy. As countries both rich and poor saw the many advantages of better-managed migration, the Global Compact for Migration became a key document in early twenty-first-century international relations and then a binding international convention in 2042.

Former colonial powers also began to revisit their historic relationships. France made a small step forward in its most remote *departement*, the island of Mayotte, a one-time colony near Madagascar that had voted to stay with France when the rest of the Comoro archipelago voted for independence in 1975. After decades of allowing Comorian boat people to drown while attempting to reach Mayotte, France offered rights of residence to Comoro citizens in 2035, though not in mainland France. The United Nations Special Committee on Decolonization worked with France, Britain, New Zealand, and the United States to grant independence to their remaining Non Self-Governing Territories (NGSTs), from Tokelau and Guam to the Falklands and New Caledonia. The climate crisis had devastated many of these small islands, so the UN Secretary General was especially pleased to announce that all residents would also be granted full dual citizenship by their former colonizers. "This is a small step toward righting the wrongs of centuries," she said in her 2040 speech marking the end of the Fifth International Decade for the Eradication of Colonialism.[17]

When Myanmar was at last held to account for the Rohingyan genocide in 2049, Bangladesh received a generous European Union aid package in exchange for its offer of residence permits to the refugees. The Rohingya would never go home again, but at last they belonged somewhere.

Projected temperature anomaly, degrees C (F)
2050–2099 as compared to 1956–2005

Part III

WARMING WATERS

What makes fever so dangerous is that the body as a whole thrives only within certain chemical and physical boundaries: its systems are set up to keep the body within those boundaries.

—Alanna Mitchell, *Sea Sick*[1]

The body of the ocean is made up of many distinct yet connected systems. In the mid-depths, which reach down a mile or more, tiny creatures play a critical role in the cycle of digestion, death, decomposition, and rebirth that links organic and inorganic matter throughout the seas. Surface waters act like a porous, skin-like membrane where oxygen, carbon dioxide, and other compounds are absorbed and released, while wind-driven currents mix the top one hundred meters of the sea and disperse organic material widely. Deep currents move volumes of nutrients slowly from one ocean basin to another over decades and centuries.[2] This exchange and flow is critical to the health of the world ocean.

Winds propel the top layer of water around the world in predictable patterns, one example being the Gulf Stream, which moves warm water from the tropical Atlantic past Europe toward the North Pole. As the warm water travels north, heat is dispersed into the atmosphere, keeping northwest Europe warmer than other regions at the same latitude: temperate Dublin, for example, lies *north* of cold Kiev. Similarly, the Humboldt Current sweeps cold water up the western coast of South America toward the equator, and exerts a cooling and drying influence on the weather of coastal Chile, Peru, and Ecuador. Farmers, navies, fishers, traders, and hunters have observed and planned their

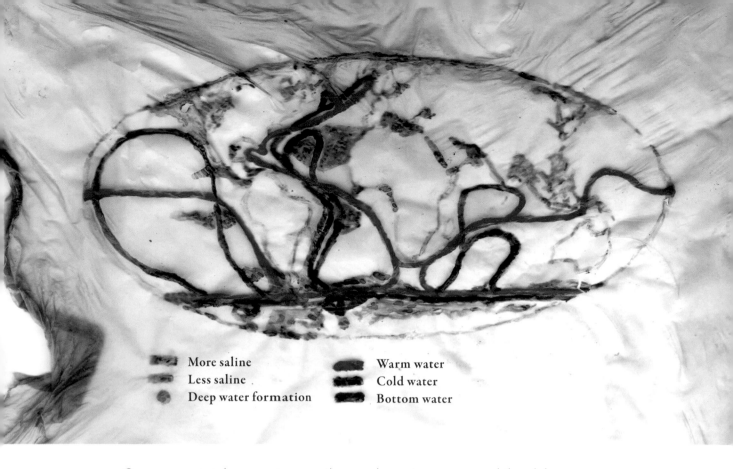

	More saline		Warm water
	Less saline		Cold water
	Deep water formation		Bottom water

Ocean currents take on water on a thousand-year journey around the globe.

activities around these currents and their effects on local environmental conditions.

As wind-driven currents move warm surface water toward the poles, this layer of water cools and grows saltier and denser through evaporation, and then sinks to join slower, deeper currents that circulate through each of the earth's oceanic basins. Scientists refer to this overall process as the Meridional Overturning Circulation, or the Ocean Conveyor Belt. In the Atlantic Ocean basin, observations of past and present conditions show that the Atlantic Meridional Overturning Circulation (AMOC) has slowed down, and the 2019 IPCC Oceans and Cryosphere report indicates that it is likely to continue slowing over the course of the twenty-first

century.[3] If this happens, Europe and northeastern North America will likely become significantly colder and wetter, leading to more snow storms and flooding. Paleoclimatologists point out that this has happened before, when the "Little Ice Age" brought crop failure and severe cold to northern Europe from 1300 to 1600 C.E.[4]

As a slow, cold current moves along the bottom of the oceanic basin, it picks up potassium, nitrogen, and phosphorus from the decomposed bodies of microorganisms on the ocean floor. Eventually this band of water, now enriched with the minerals that are necessary for plant growth, collides with a continental shelf and is forced back up to the surface in a process known as upwelling.

Here, nutrients mix with sunlight and dissolved carbon dioxide, making possible the massive phytoplankton blooms on which all marine ecosystems depend. This surface layer is then driven by wind currents back toward the equator, and the cycle repeats. Through this Ocean Conveyor Belt, a water molecule that was once frozen on the tail feather of an Antarctic penguin may find itself drifting across the eye of a Kenyan parrotfish a few hundred years later, and blasting from the blowhole of an Arctic narwhal several hundred years after that.

Along the Ocean Conveyor Belt are five major gyres: the North and South Pacific, the North and South Atlantic, and the Indian Ocean. Each gyre is made up of several currents that circulate (clockwise in the Northern Hemisphere and counterclockwise in the Southern Hemisphere). Three factors create gyres: the rotation of the earth, which spins the planet itself and also the water on it; the sun, which heats the atmosphere and creates wind; and the continents, which act as boundaries to the wind and water. To create your own tiny gyre complete with different water densities, pour a splash of cream into your coffee and stir. The liquid will rotate in the direction it was pushed by the spoon within the cup. The cream first sinks because it is denser than coffee, but some rises again through mixing. If you've made your gyre in a clear glass, you can observe the different layers of light and dark liquid and how they mix together.

The relative strength and weakness of the currents within each gyre have major weather impacts on land and in the seas. For example, when the cold Humboldt Current weakens in the South Pacific Gyre, warm water from the eastern Pacific is pulled farther west. This is known as the El Niño phase of the Southern Oscillation (ENSO). The El Niño phase happens every two to seven years, and historically the more extreme phases have happened once every twenty years. During extreme phases, impacts are felt around the globe: deadly mudslides in California, fisheries collapse in Peru, cholera in Tanzania, dengue fever in Southeast Asia, and crippling drought in Australia's eastern plains.[5] As waters continue to warm, researchers believe that extreme El Niño events will happen more frequently.[6]

Over the past fifty years, the ocean has absorbed 90 percent of the heat we have added to the climate system. According to NASA, "while the atmosphere has been spared from the full extent of global warming for now, heat already stored in the ocean will eventually be released, committing Earth to additional warming in the future."[7] And the ocean is warming at a rate 40 percent more quickly than the IPCC estimated just five years ago.[8] This heat, though measurable in fractions of degrees, dramatically changes the metabolic and immune processes in marine creatures, from unicellular plankton, the basis of the oceanic food web, to coral polyps, the architects of reef ecosystems. Researchers think that warming may also contribute to increased ocean stratification, a situation in which the layers of ocean water mix less efficiently. As warmer air and water melt polar ice at a faster rate, this creates a less salty, and therefore lighter, surface layer, which doesn't sink as easily. If this surface layer doesn't sink, the entire Oceanic Conveyor Belt could

THE ARCTIC OCEAN: WHEN THE ICE MELTS

> One could say Indigenous peoples are ahead of the curve in terms of processing the emotional shock and trauma of cataclysmic environmental and social devastation.
>
> —Kyle Whyte, Indigenous scholar[1]

Sea ice first forms as tiny, cellophane-thin, star-shaped crystals that spread across the water on calm, cold nights and merge into a fragile shell on the sea. Wind and waves break the surface into shards that recombine, forming a kind of slush called frazil or "grease ice." As freezing temperatures bind loose pieces of frazil into larger sheets, the lumps of ice bump and jostle against each other, rounding their sharp edges into thicker ovoid forms called pancake ice. Sea ice can grow up to twenty centimeters (eight inches) thick above the surface, at which point it continues to grow from below the surface, reaching a depth of up to 1.5 meters (five feet), with longer columns stretching far deeper. This first-year ice is the foundation of the Arctic system, accumulating across the northern ocean every winter to form a polar ice cap that is constantly fracturing and reconfiguring, always on the move.

If sea ice stays frozen for two years or more, it is called multi-year ice, and it can grow up to five meters (sixteen feet) thick, spreading to the horizon in every direction. Multi-year ice is critical to the Arctic ecosystem because it forms the cooling lid that prevents the ocean from warming too much each summer and over time. Without it, the planet's air-conditioning system would fail for half the year, allowing the global climate to fluctuate far more dramatically. But the Arctic air and sea are heating faster than any other region of the planet: 2019 was the second-warmest year on record, with sea temperatures up to 4°C (7.2°F) warmer than

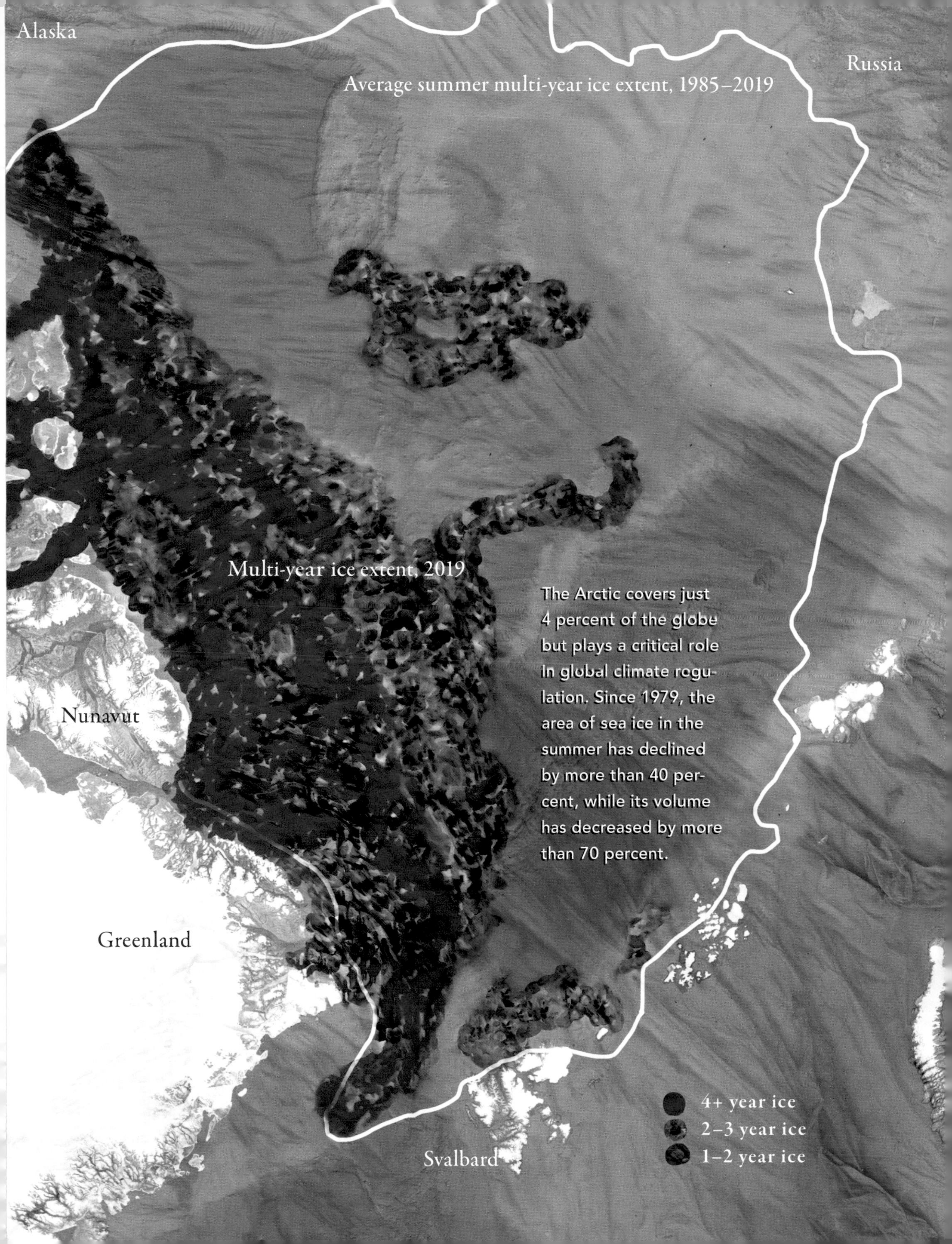

Alaska

Russia

Average summer multi-year ice extent, 1985–2019

Multi-year ice extent, 2019

Nunavut

The Arctic covers just 4 percent of the globe but plays a critical role in global climate regulation. Since 1979, the area of sea ice in the summer has declined by more than 40 percent, while its volume has decreased by more than 70 percent.

Greenland

Svalbard

4+ year ice
2–3 year ice
1–2 year ice

I. OPEN WATER

$\alpha = 0.06$

1

0.94

II. BARE ICE

$\alpha = 0.5$

1

0.5

III. ICE WITH SNOW

$\alpha = 0.9$

1

0.1

Albedo, or the measure of the reflectivity of surfaces, inverts as the Arctic ice cap melts, speeding global warming when a blue ocean absorbs almost all the sun's heat.

eighty-four times more potent than carbon dioxide over a twenty-year time frame. If released in a short time span by sudden warming, underwater methane releases could quickly bounce the average global temperature up by another 0.6°C (1°F), with catastrophic consequences. But little methane research and almost no monitoring is being done.[9]

Other positive feedbacks pulsing through the Arctic include increased wave energy, which breaks up more ice faster; increased water evaporation, which acts as a heat-trapping greenhouse gas; and water pooling on melted ice, which causes the underlying ice to melt even faster. "We are not far from the moment when the feedbacks themselves will be driving the change, that is, we will not need to add CO_2 to the atmosphere at all, but will get the warming anyway," warns Wadhams.[10]

We do not know if these things are certain to happen, because the data is relatively sparse, but as Gavin Schmidt, a NASA climate scientist, notes, "The planet is very capable of surprising us."[11] But as measured observations accumulate, many scientists have revised their most recent projections: the Arctic could be ice-free in September in little more than a decade, and throughout the summer by the mid-2030s.[12]

∼∼∼∼∼∼∼∼∼∼∼∼∼∼∼

Key Term: Vulnerability

The word *vulnerability* was rarely used until the 1950s, but it has since leapt into com-

emitting tens to hundreds of thousands of gigatons of carbon that cannot be re-stored.[8]

Even more disruptive may be the melting permafrost that lies just below sea level. The continental shelf of eastern Siberia, which was above the waterline during earlier ice ages, is now a drowned, carbon-rich tundra kept frozen by the cold water and seasonal ice above it. As the water heats, these shallow underwater lands are decaying and releasing methane hydrate, a greenhouse gas mon speech with the rise of both popular psychology and scientific risk analysis. Environmental scientists measure vulnerability to assess the viability of species and ecosystems, and the medical community conducts similar analyses on people and populations. In psychology, vulnerability has been used sometimes synonymously with weakness; we don't want to be reminded of our fragility, impermanence, and, ultimately, death. Yet there are other ways of seeing vulnerability not as a flaw to guard against, but as an inherent quality of complex systems. Some religions and schools of thought include practices that cultivate acceptance of change and the discomfort it may cause in order to build inner strength. In so doing, vulnerability becomes a way of relating to, rather than feeling alienated from, other people and the living world.

A View from 2050

Ever since the Swedish scientist Svante Arrhenius first identified the "greenhouse gas effect" in 1908, northern countries have been eyeing the advantages of a warming Arctic.[13] By 2020, Russia was using its fleet of icebreakers to guide freighters across its Northern Sea Route, where multi-year ice had become rare. China was investing heavily in Russian gas projects to secure both future energy supplies and a faster route to European markets, while many nations with nominal claims, from the United Kingdom to India, were jockeying for position in the region. Russia, Canada, and Denmark had

applied to have part or all of the Lomonosov Ridge—the shallow underwater mountains that bisect the entire Arctic Ocean—added to their continental shelf territorial claims, along with all the associated drilling rights. With an estimated 30 percent of the world's undiscovered oil and gas under the melting ice cap, a balance of geopolitical power as delicate as the Arctic ecosystem blanketed the region.[14] The ice was melting so fast, and there was so much money to be made.

Military and mining operations grew tense and sometimes confrontational in the 2020s, yet Arctic countries refused to develop any binding agreements for fear of limiting their own claims. The Northwest Passage, the holy grail of sea exploration along the Canadian and Alaskan coast, became a reliable shipping route, bringing significant income ashore in Iqaluit, the capital of Nunavut. Canada's main planning document, the Arctic Policy Framework, was soon hopelessly out of touch with climatic and cultural realities, and Indigenous communities used the phrase "genocide by bureaucracy" to describe Ottawa's inaction as summer temperatures surged. Eventually, the government of Nunavut negotiated the right to draft the region's next policy plan. *Arctic 2050* addressed the need for greater income and independence with a stiff tourist tax that would be reinvested in a "blue economy" of biofuel farms, wind energy, and tourist amenities to increase the region's self-sufficiency.[15] "Feedback" became the word of the decade, as people below the Arctic Circle began to understand that they needed to flatten the curve of climate change before the changes overtook them.

Panama Current

Galapagos Islands

Sea surface temperature anomalies
are far greater near coastal Peru
during extreme ENSO years, whereas
average ENSO years have greater
effects in the central Pacific.

Ecuador

Peru

Peru Oceanic Current

anchovy
fishing
boundary

Humboldt Current

Sea surface temperature anomalies, degrees Celsius

+.3   +.5   +.7   +.9

Pisc

PISCO, PERU: ENSO AND THE END OF FISH

It was a lone voice in the middle of the ocean, but it was heard at great depth and great distance.

—Gabriel García Márquez, *Love in the Time of Cholera*[1]

A rainbow of wooden boats bobs in the Pisco harbor, the artisanal fishing fleet at rest from its pursuit of the Peruvian *anchoveta*. Known more formally as *Engraulis ringens*, this member of the anchovy family surrenders an average of five million metric tons of its flesh to the fishing industry every year. The Peruvian anchovy fishery is the biggest fishery in the world, and the town of Pisco (also the birthplace of the eponymous sour) is one place where anchovies transition from being ocean wildlife to global commodity. The slender, glittering *anchoveta* is at the center of a globalized web of trade that links pigs in rural China to bond traders in New York and is a mainstay ingredient of both subsistence shrimp farm feed in the Philippines, and omega-3 prenatal smoothie boosts in Los Angeles.

E. ringens, like their anchovy and sardine cousins throughout the world, are found primarily along the eastern edges of oceans, in the cold, nutrient-rich waters of eastern boundary currents. The Peruvian anchovy is about 14 centimeters (5.5 inches) long—the length of a child's hand from fingertip to wrist. It spends its entire life feeding, procreating, and dying in the open water column of the Humboldt Current. This is one of the largest upwelling systems in the world, flowing in a nine-hundred-kilometer-wide band up the South

THE NORTH ATLANTIC: IN DEEP

All flesh is grass.

—Isaiah 40:6

The diversity of microscopic life in the sea is almost unfathomable, an infinite web of bacteria, protists, viruses, and tiny animals that are together called plankton, "those who are made to wander" in Greek. These smallest of sea creatures are highly sensitive to changes in seawater, some even more so than fish or marine mammals. Changes to the ocean's temperature alter their microbiology, which in turn change the biogeochemical cycles of the ocean itself, cycles that govern many of our planet's fundamental processes.[1] Little things mean a lot in the life of the sea.

Plankton animate each liter of salt water with billions of tiny lives, yet these creatures are among the least recognized forms of life on earth. By one estimate we have identified just 9 percent of the 2.2 million species in the sea.[2] Most important are the phytoplankton, the (usually) single-celled creatures that perform the "vital alchemy of photosynthesis," consuming carbon dioxide and sunlight and creating energy and oxygen.[3] Their respiration generates 50 to 80 percent of the oxygen on earth, and alongside the tropical and temperate rainforests, they form the other great lung of our planet.[4] But they are not plants, and their structures and functions are uniquely suited to the sea. Commonly called algae, the phytoplankton include wildly different creatures, from single-celled, glass-shelled diatoms to giant forests of twenty-foot-long bull kelp, and researchers who study the socio-ecology of marine microbes argue that phytoplankton "may be an intermediate state between single cells and aggregates of physically attached cells that communicate and co-operate; perhaps an evolutionary snapshot toward multicellularity."[5]

The most abundant phytoplankton are the photosynthesizing bacteria, also called cyanobacteria for their blue-green color, that evolved more than three billion years ago. The earliest varieties, thought to be the first life on earth, completely transformed the

*2018 Sea Surface
Temperature Anomaly*

Greenland

Canada

Sea surface temperature
in the North Atlantic is
rising faster than almost
anywhere on earth.

Canada

USA

earth's climate with their respiration process from an oxygen-free environment to the oxygen-rich atmosphere the world has known ever since.[6] This was an environmental crisis for all life at the time. The only life-forms that survived this "Great Oxygenation" crisis were those that evolved to process oxygen, the toxic and highly combustible gas we now breathe. Those that didn't adapt perished, causing the first mass extinction. Without cyanobacteria, life would have evolved down a very different path, or not at all.[7] Yet we know little about them. One genus within the cyanobacteria phylum, called *Prochlorococcus*, is the most productive energy generator in the ocean and is perhaps even the most abundant form of life on earth, with 100,000 cells in each milliliter, or twenty drops, of seawater. Because it is so small (0.6 micrometers), it was only discovered using electron microscopes in 1986.[8]

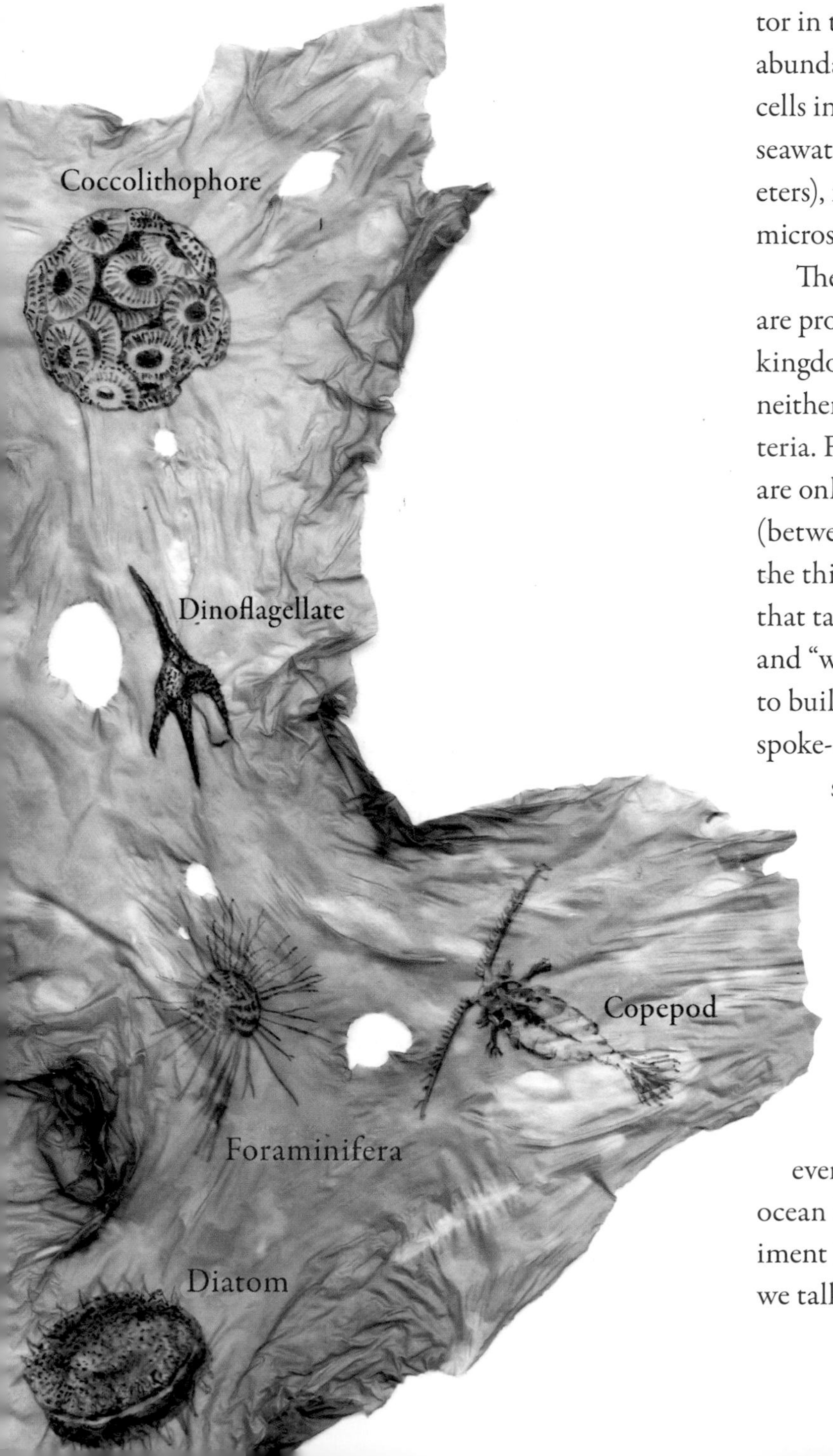

The rest of the ocean's photosynthesizers are protists, members of a vast taxonomic kingdom of single-celled organisms that are neither plant nor animal nor fungus nor bacteria. For instance, coccolithophores, which are only 0.1 to 1 millimeter in diameter (between the diameter of a red blood cell and the thickness of a thumbnail), are protists that take molecules of calcium carbonate and "weld them with the torch of sunlight" to build intricate, interlocking, wheel-and-spoke-shaped calcite plates called coccoliths, some thirty per creature.[9] When predators eat these plankton, they excrete the non-nutritious plates, which are usually digested back into dissolved organic material by bacteria. This "endless cycle by which the ocean consumes and recomposes itself" is called the biological pump.[10] Some of the calcium-rich coccoliths, however, don't dissolve, instead sinking to the ocean floor where they stay locked in the sediment for centuries or even millennia. When we talk about the ocean as a carbon sink, we

Coccolithophore

Dinoflagellate

Copepod

Foraminifera

Diatom

Increase in jellyfish population size

\ High certainty   / Medium certainty | Stable   / Decrease

Photosynthesis in the ocean is more pronounced at the poles and is dependent on dissolved micronutrients that flow with the currents. The majority of large marine ecosystems have experienced a 60 percent increase in jellyfish numbers due to warmer waters.

are mostly talking about coccolithophores, which alone sequester half of all the carbon in the sea.[11]

Equally bizarre and beautiful are the diatoms, photosynthesizing protists that build glass houses for themselves out of silica, each in the shape of a tiny pill box patterned with intricate ridges that distinguish the twenty thousand to thirty thousand species. In the North Atlantic, when the days get longer in the spring, diatoms fill the cold waters, doubling in number daily but living just a few hours.[12] All marine photosynthesizers live in these brief, spectacular bursts called

algal blooms, and each population explosion triggers a banquet as the zooplankton—the slightly larger, multicellular animals—come up from deeper waters to feed.

The zooplankton are dominated by copepods, tear-drop-shaped crustaceans that hibernate all winter in the deep ocean, waking just in time for the feeding frenzy. They, in turn, are a major food source for larger animals, from krill to whales. For instance, the North Atlantic copepod *Calanus finmarchicus*, which is the size of a grain of rice, is the main food source for cod and herring larvae and is therefore an essential piece of the North Atlantic ecosystem and economy. This seasonal dance between predator and prey takes place at a tight tempo, but if the algae bloom before the copepods wake from their winter slumber, the predators won't have enough food.[13] These species can recover from normal climate variability—an early spring one year, a long, cold winter the next—but prolonged warming causes systemic change, as it is, in fact, already doing. The collapse of the cod fishery in the North Atlantic in the 1980s and '90s was the result of a combination of these warming waters and overfishing. Warmer water is likely why the cod stocks in the North Atlantic took decades to recover from their collapse, even after severe catch restrictions were implemented.

Slowly but surely, we are beginning to learn about the biorhythms of the ocean. Each night, as the sky darkens, the largest migration on earth takes place as plankton move en masse from the mid-ocean depths to feed nearer the surface. Moving as much as fifty thousand times their body length in a just few hours, trillions of living creatures make their way up through the water column using flagella, air sacs, lipid bodies, and cilia, then sink down again in the morning to avoid the sun's damaging rays. These suspended bodies amount to 50 to 80 percent of the life mass on earth. Yet, as Heidi Sosik, an oceanographer at Woods Hole Oceanographic Institution, says, "We don't know which species are migrating, what they're finding to eat, who is trying to eat them, or how much carbon they're able to transport."[14]

The dark ocean below 200 meters, long thought to be uninhabitable, is a very cold but very stable zone of small, sophisticated creatures. For instance, ctenophores, also called sea gooseberries for their size and appearance, voraciously hunt copepods. Tomopteris, a centimeter-long, luminescent, segmented worm with two dozen flippers, eats most anything, and the bristlemouth, a toothy, transparent fish just one centimeter long, has recently been found to be the most abundant vertebrate on earth. Deeper still live the archaea, an entire domain of single-celled life that was only discovered in the 1970s and which early research suggests will thrive in a warming ocean. With an average depth of 4,000 meters (2.5 miles), the deep ocean and its bed of earth are home to 40 percent of all marine microbes. Yet it is one of the least studied ecosystems on earth. "There are places in the deep ocean that, in thirty years, will never again be the same temperature as they are now, and we don't know what lives there, and we don't know what will happen to them," according to Alyson Santoro, professor of ecology at the University of California, Santa Barbara.[15]

Foraminifera, protists that range from 100 micrometers to 20 centimeters (up to 8 inches) in size, thrive in the benthic mud, building ornate calcite shells, out of which they send long, tentacular pseudopods to hunt. Bacteria, too, abound deep in the oxygen-free zone. And marine viruses reside within every single marine creature—at least 200,000 species and 10^{30} in number, or more than the number of stars that exist in the visible universe. Not exactly alive, but certainly not dead, viruses play critical roles in ocean biology that have yet to be thoroughly studied. Scientists believe bacteria and viruses will have an evolutionary advantage in a warming ocean, because they may be able to adapt more quickly than complex life-forms.[16]

Data from the Continuous Plankton Recorder (CPR), which has been tracking transoceanic plankton populations since 1931, show many changes already rippling through the microscopic food web. Warmer-water plankton species are moving northward into the cooler North Atlantic at a rate of about twenty-three kilometers (fourteen miles) per year, significant shifts that "are similar in size to large marine ecosystems and state exclusive economic zones." The cold-water species are, in turn, migrating to even higher latitudes, altering the food web in ways we do not yet understand. CPR researchers have recently identified a new plankton species in the North Atlantic that is the first documented instance of trans-Arctic species migration.[17] Though largely invisible to the naked eye, the oceanic food chain is rapidly transforming, and sometimes disappearing, on our watch.

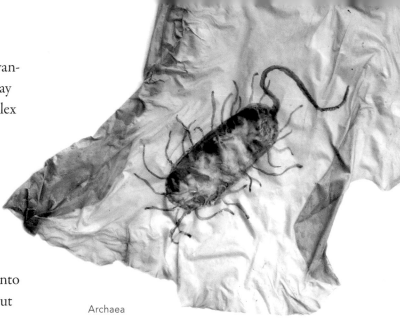

Archaea

Key Term: System

In the mid-twentieth century, *systems theory* grew out of cybernetics, the study of complexity and communications, which argued against mechanistic models that reduce matter to mere components. Systems thinking recognizes that the relationships and interactions between a system's parts are as important as the parts themselves. As such, a system—a galaxy, say, or a seal—is greater than the sum of its parts. Systems regulate themselves through feedback loops, which allow them to maintain homeostasis while also adapting to change. By definition, then, systems are both whole in themselves and continually learning, changing, and emerging. They are both process and product, complex and irreducible, organized yet permeable, dynamic and generative. For a number of twentieth- and twenty-first-century thinkers, this theory has opened the sciences to new avenues of inquiry.[18] In the words of mathematician Norbert Wiener, founder of cybernetic theory, "We are not stuff that abides. We are patterns that perpetuate themselves; we are whirlpools in a river of ever-flowing water."[19]

A View from 2050

In the early 2000s, techniques developed by ocean ecologists—including metagenomics and single-cell sequencing—revolutionized the biological sciences, making possible the sequencing of the human genome. Such multidisciplinary collaborations encouraged marine scientists to reach out to cognitive scientists, medical researchers, and even artists to expand their inquiries, breaking through barriers that had long divided their fields. By the 2010s and 2020s, teams were exploring cognitive theory in microbial species, applying artificial intelligence models to plankton to ask whether microbes could learn through practice, and theorizing system-level learning in marine environments.[20] By 2025 the Ocean Memory Project had radically changed our understanding of microbial learning and memory, and gradually this discovery had the Copernican effect of moving humans away from the center of the biological universe.[21] It was a turning point for academic work in evolutionary science, philosophy, and psychology, which trickled into the mainstream for decades.

Through the 2020s and 2030s, ecologists and regulators were in a race against factory ships, which had begun industrial fishing at ever deeper levels, devastating the "twilight zone" two hundred to one thousand meters below the surface. Trawling and dragnetting tore apart food webs, even as scientists in the U.S. National Microbiome Initiative worked against the clock to describe and study as many species as possible.[22] This well-funded multinational effort became the starting point for many medical breakthroughs, including effective therapies for Parkinson's disease. The relatively new field of marine bacteriology and virology became a new focus of drug research in a bid to find medicines to beat the antibiotic-resistant

superbugs and pandemic viruses that were emerging ever more frequently.[23] But we will never know if any of the species we failed to save in the first half of the century might have eventually saved us.

In 2050, the North Atlantic sea surface temperature is now 2 to 4°C warmer than it was at the beginning of the century, depending on location, and oceanographers' earlier predictions of rolling extinctions have proven largely accurate. Complex connections and feedbacks have countermanded some earlier predictions and amplified others, with researchers barely able to keep up.[24] For instance, it was predicted that cod and herring populations would decline and haddock would benefit in response to changing plankton profiles. Instead, haddock have also collapsed as less organic material makes its way to their seafloor feeding grounds.

Norway, the world's commercial aquaculture powerhouse, invested early on in fish farms, which has paid off generously. When their offshore waters became too warm for Atlantic salmon, they built large onshore facilities fed with geothermally cooled seawater. The fish ate feed made from farmed insects rather than wild fish, a more cost-effective and environmentally sensitive choice. Salmon, once Norway's second-largest industry, recently replaced oil and gas as the country's main revenue source.

With fewer predatory fish in the ocean, jellyfish, the largest plankton of all, have filled the gap in the North Atlantic food web. A low metabolism helps them survive in warmer water, making them better hunters than their fishy competitors, who become slow in low-oxygen environments.[25] Tropical species migrated north when their native ecosystems became too hot, drifting into fish farms and onto northern European beaches. Blue blubbers, bushy bottoms, fire jellies, jimbles, cannonballs, sea walnuts, pink meanies, hair jellies (or snotties), mauve stingers, and Portuguese man-of-war now inundate open waters and coastlines all over the world.[26] Some are edible—delicacies even—but most wash ashore in slimy masses, making "jelly-day" beach closures miserably common and full-body swimsuits necessary in many places. From fishing nets in Japan to power plant intake valves in Israel to aquaculture pens in Ireland, jellyfish are drifting their way toward dominating the seas of the twenty-second century.[27]

KISITE, KENYA: CORAL COLLAPSE

I have the utmost respect for corals, because I think they've got us all fooled. Simplicity on the outside doesn't mean simplicity on the inside.

—Ruth Gates, *Chasing Coral*[1]

Coral is easy to love to death. In the Kisite-Mpunguti Marine National Park off the southern coast of Kenya, tourists scrape their sunburned knees and bellies over a bed of coral skeletons. Their skin and sunscreen slough off into the water, coating the coral in chemicals that protect the tourists from getting further burned, but that also prevent the coral polyps from being able to photosynthesize their food. Starvation-by-sunscreen is not the only way that corals are dying these days; rising ocean temperatures are also killing these biodiversity hotspots.[2]

Corals defy our assumptions about what it means to be alive. They are a weird hybrid of photosynthesizing algae and poison-wielding carnivore. They are the builders of, and also the keystone species in, reef ecosystems. *Keystone*, a loaner word from the field of architecture, is a particularly apt word in the context of reefs, where the corals provide the physical foundation upon which daily life unfolds: spaces for critters to live, rest, procreate, hide, and feast. So when corals are in bad shape, the entire ecosystem is in trouble. And that usually means that the well-being of the people who depend on the reef is also at risk.

Tropical coral reefs are made up of magnitudes of tiny creatures living in symbiotic colonies. The carnivorous animal part of coral, called a polyp, has a sac-like body and stinging tentacles surrounding its mouth.

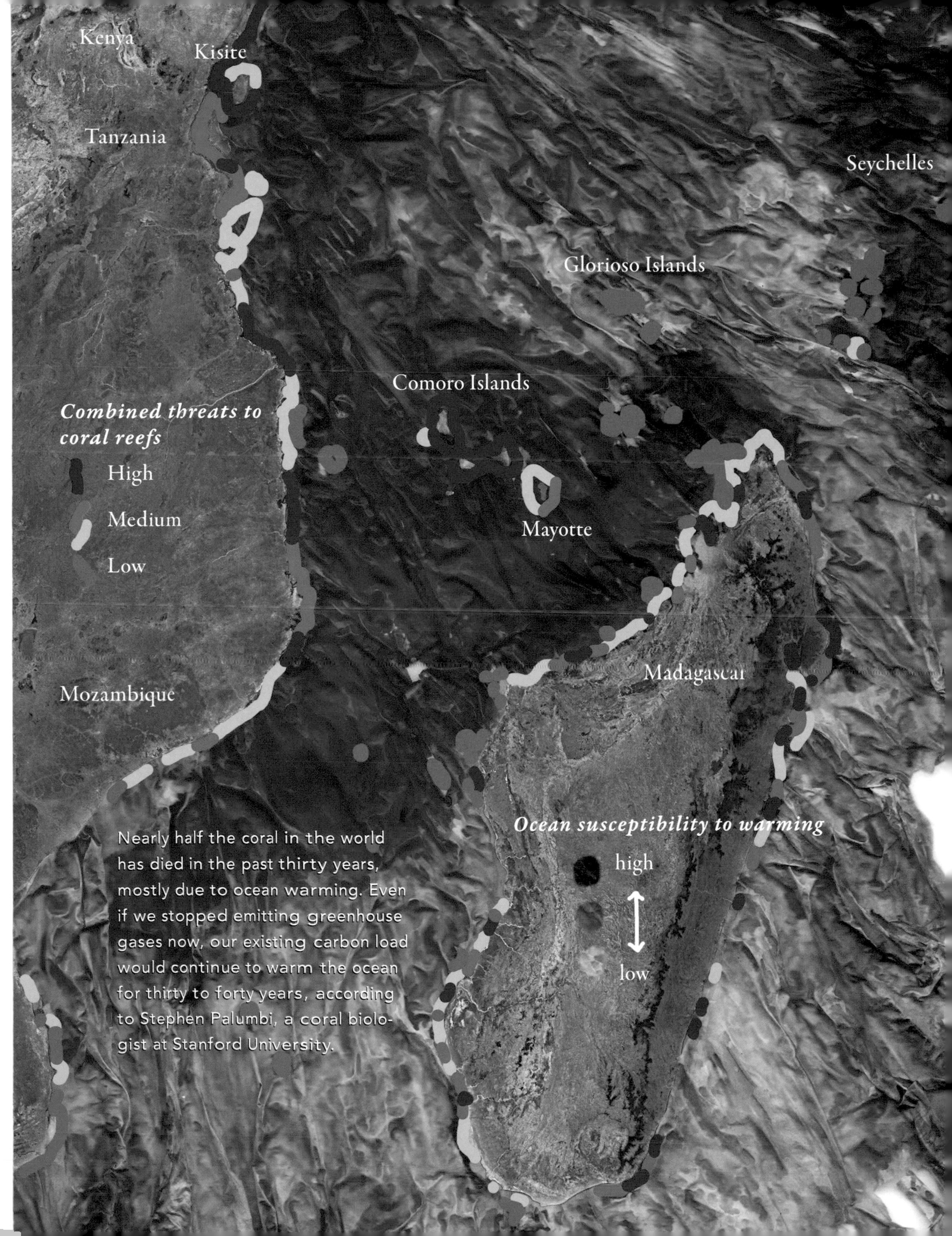

Kenya

Kisite

Tanzania

Seychelles

Glorioso Islands

Comoro Islands

Combined threats to coral reefs

High

Medium

Low

Mayotte

Mozambique

Madagascar

Ocean susceptibility to warming

high

low

Nearly half the coral in the world has died in the past thirty years, mostly due to ocean warming. Even if we stopped emitting greenhouse gases now, our existing carbon load would continue to warm the ocean for thirty to forty years, according to Stephen Palumbi, a coral biologist at Stanford University.

conditions that are a direct result of our actions.[10] In 2019, scientists at the Australian Institute of Marine Science introduced the first assisted-evolution "supercoral" babies onto the Great Barrier Reef.

When it comes to surviving the Anthropocene, we are the problem and must also be the solution.

~~~~~~~~~~

Key Term: Evolution

Charles Darwin laid out his theory of *evolution* in 1859 in *On the Origin of Species*. He provided evidence that plants and animals respond to their environment by changing slowly over generations to compete more effectively. Although we sometimes think of our understanding of evolution as starting with him, humans actually had our finger in the evolutionary pie for millennia before Darwin wrested the world of biology from the hands of the church. Part of our common human heritage is plant and animal husbandry. For better and for worse, it is our tinkering and tampering with nature—including our "assistance" with the process of evolution—that has made us the dominant species on our planet. Our power to manipulate the world around us has grown beyond the husbandry of yesteryear to a point where fish genes can be spliced into a tomato to extend its shelf life. Helping ocean species to survive climate change will further increase our footprint on the world and force us to wrestle with the ethical considerations of having so much power.

~~~~~~~~~~

A View from 2050

Wild coral reefs are now nearly gone. A combination of rising temperatures, unsustainable fishing practices, over-tourism, deoxygenation, and ocean acidification pushed many coral reef ecosystems to a tipping point in the late 2020s and early 2030s from which they never returned. A typical reef now might have 10 to 15 percent of its coral still alive, but this isn't enough for the ecosystem to recover and keep pace with climate change. The reefs that persist now do so through the direct intervention of people. This domestication of coral got under way in the 2020s once it was clear that reefs weren't going to make it on their own. Scientists, managers, and local residents have propagated three very different kinds of reefs based on the services that they provide to people: tourist gardens, community farms, and wildlife sanctuaries. The Kisite-Mpunguti Marine Conservation Area complex is a model of how to make space for all three.

Most of the eco-resorts in Shimoni and Diani Beach (where visitors to Kisite stay) have a coral garden adjacent to the hotel. These pretty patches of colorful corals are reached by a well-groomed wading path, and their perimeter is marked by buoys so that visitors won't stray into an off-limits community farm or a wildlife sanctuary. Hotels treat the corals in these gardens like annual plantings—easily replaced eye candy that bolsters the resort's reputation for good photo ops and fun snorkeling. If the water in these gardens gets too warm and

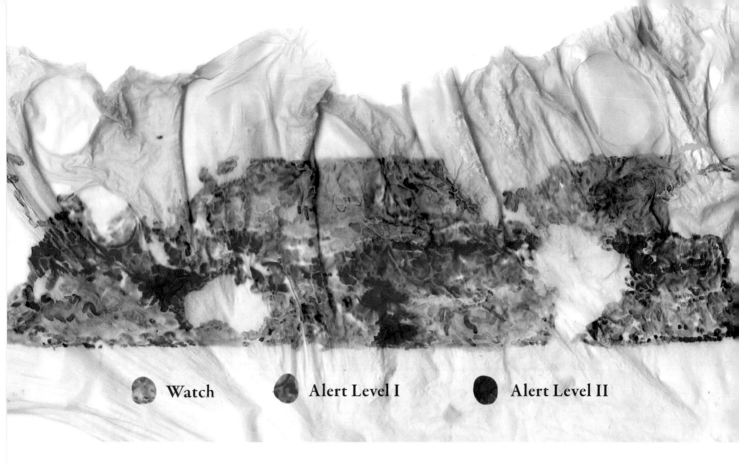

Watch ● **Alert Level I** ● **Alert Level II**

Map of coral stress, August 5, 2018

the corals die, the hotel just ships in more specimens in the most attention-grabbing colors and shapes. Little attention is paid to whether the replacements are native, cross-bred, or transplanted from another ocean. A well-maintained adjacent reef is now among the standard set of amenities for the pricier hotels, along with beach yoga, SUP rentals, and boat excursions to Shimoni's British colonial ruins.

By the mid-2030s, the global-brand hotels had figured out a solid formula for feel-good tourism on the Kenyan coast. First, follow the tried-and-true strategy of buying out the existing lodging and ecotour operations from the locals and then consolidate these hold-ings into all-inclusive resorts. Next, fund a coral reef restoration project adjacent to the hotel for easy guest access (bonus points if the restoration work can be connected with a well-respected research institution). Finally, get the global brand's sustainability director, corporate brand manager, and communica-tions people to craft a tale of good corporate citizenship that social media influencers would carry to the masses.

The coral farms are fairly utilitarian and also fairly well patrolled by locals, who had their rights and responsibilities returned in the 2030s to tend and defend the reefs as an important food source. These areas are spe-cifically set aside to harvest the animals that

PINE ISLAND GLACIER: WHAT HAPPENS IN ANTARCTICA DOESN'T STAY IN ANTARCTICA

> Vast uninhabited wastes are much easier to share than a billion barrels of oil or veins plump with ore, and neither the U.S. nor any other nation wants to be left in the lurch when the Antarctic bauble has finally been appraised.
>
> —Nicholas Johnson, *Big Dead Place*[1]

While most people will never set foot on Pine Island Glacier (or PIG, as many researchers refer to it), the disappearance of this Antarctic glacier will be experienced around the world through sea-level rise and changes to weather patterns. The problem is, no one knows how quickly PIG is disappearing or how deeply we will be touched by its loss. This is the source of the greatest uncertainty for global sea-level rise projections.

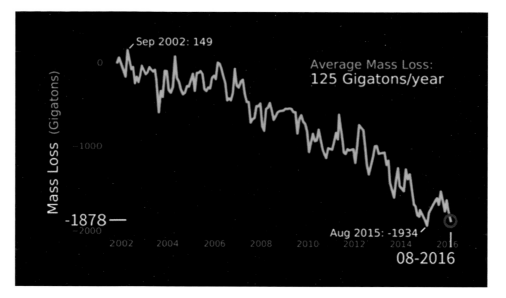

As the air and water warm, Antarctic ice is melting at a faster rate.

tions) have melted enough to make natural resource exploitation feasible.

Key Term: Uncertainty

Physicist Richard Feynman said, "It is imperative in science to doubt; it is absolutely necessary, for progress in science, to have uncertainty as a fundamental part of your inner nature. … The statements of science are not of what is true and what is not true, but statements of what is known to different degrees of certainty."[8] The Intergovernmental Panel on Climate Change (IPCC) has struggled for decades with how to communicate scientific *uncertainty* to an inattentive public, impatient policymakers, and climate change skeptics.

Since 2010, the IPCC has used a very specific formula to designate the levels of confidence and uncertainty for each of their projections. Each projection is evaluated by the amount and quality of existing evidence, as well as the degree of agreement among experts. Confidence is expressed as a qualitative judgment about the validity of each finding—a high level of robust evidence and high agreement of experts results in "high confidence," while lower levels of agreement or evidence lower the confidence. There is no measure of absolute certainty, and the scale only goes up to 99 percent, or "virtually certain." In times of disaster—financial, societal, environmental—we want our leaders to act decisively to protect us from harm. Producing accurate, credible information that leaders need requires a tolerance for mistakes and continued willingness to engage in the messy public process of scientific deliberation. The deliberative pace of science does not align well with that of a 24/7 news cycle, and the research culture of embracing doubt doesn't sit well with politicians or a public seeking decisive action. But the coming decades may force us all to become a bit more tolerant of the discomfort of uncertainty.

A View from 2050

On the first day of the Southern Hemisphere's summer, December 21, 2050, the new International South Pole Visitor Station opened. The digital marquee at the entrance of the main building displayed the greeting "Antarctica, Open for Research, Open for Exploration!" in French, Spanish, Portuguese, Norwegian, Hindi, English, and Chinese. Scientists, politicians, and tourists were here to celebrate the extension of the Antarctic Treaty for another thirty years. The "Open for Exploration" part of the slogan left some feeling queasy, wondering if "exploration" would remain limited to tourists marveling at icebergs, or would eventually include corporations poking and prodding to locate natural resources. But at least for the moment, Antarctica was still closed to mining and military exercises.

The Antarctic Treaty had seen many ups and downs throughout the early twenty-first century. In the 2020s, the scientific community had begun speaking frequently and forcefully about how climate change impacts in Antarctica could spread around the world. By 2045, the edges of the Thwaites and Pine Island glaciers had been chewed away to the point that the West Antarctic Ice Sheet was destabilizing. Now, at mid-century, scientists are virtually certain that at least six of the nine inches of sea-level rise we've had since the turn of the century are due to Antarctic meltwater. There is also consensus that the collapse in Southern Hemisphere fisheries is the result of the slowing of the southern part of the Ocean Conveyor Belt.

In 2040, South Africa, one of the original Antarctic Treaty signatories, announced a research program to explore the feasibility of harvesting freshwater icebergs. The announcement about the feasibility research was made on the same day that a ship in South Africa's presumed exclusive economic zone waters began pushing a small berg into the current that would float it up to water-starved Cape Town. It was clear that this was not an exploration of theoretical ideas, but rather the execution of an existing plan to materially benefit one of the signatories. Argentina argued that this violated the spirit of the existing Antarctic Treaty and accused South Africa of conducting extractive activity without the appropriate environmental review or permits. South Africa responded that it was a proof-of-concept exercise that had the potential to benefit other water-starved regions as well. Once the first iceberg successfully reached and watered Cape Town, Argentina threatened to pull out of the treaty altogether unless the activities were stopped. South Africa agreed to halt iceberg harvesting for a year, during which time the Antarctic Consulting Parties meetings were the most lively they'd been since the treaty was first formed, as the parties argued over what activities could be considered public-good research, who was permitted to conduct such research in unclaimed lands and waters, and what the consequences of breaking the treaty would be.

A big slice of PIG stopped the bickering. An iceberg 500 square kilometers (193 square miles) heaved into the sea with little warning in 2046, and over the next six months, several smaller bergy bits totaling another 100 square kilometers (39 square miles) dribbled into the Amundsen Sea. The fear of failed negotiations and a resulting free-for-all hung over the treaty renewal negotiations, so the parties opted to keep most of the existing limitations on activities for another ten years. This would ostensibly give scientists a little more time to observe what was happening with PIG and what its continued disintegration might mean for sea levels around the world. The biggest change to the interim treaty was to add ecotourism explicitly to the very short list of allowable activities. The stated rationale for such an odd policy move with so many uncertain consequences was to increase the public's scientific literacy and awareness of what was at stake by allowing more people to experience the region. The updated treaty wasn't the complete victory the precautionary principle activists had hoped for, and researchers weren't happy about the prospect of tourists underfoot, but it was better than outright exploitation.

Part IV

RISING SEAS

Sea level rise is like an invisible tsunami, building force while we do almost nothing.

—Ben Strauss, Climate Central[1]

Swelling, sweating, spilling over, and sloshing around where it shouldn't, sea level is our clearest sign yet that the fossil fuel party is over. All along our coastlines, we are starting to see the long-term consequences of short-sighted decisions, from drowning wetlands to sunny-day flooding in cities. The analogy between human bodies and the ocean's body—between systemic inflammatory illnesses like diabetes and obesity and the permanent damage caused to shoreline communities by rising seas—is clear. Our globe's got gout.

About 40 percent of humanity currently lives within one hundred miles of the sea. This puts us at risk of losing thousands of communities and trillions of dollars' worth of desalination plants, boardwalk amusement parks, wastewater treatment facilities, beach houses, office towers, airports, roads, and more. We may soon lose pieces of our cultural heritage, and important archaeological sites, including Skara Brae, Scotland, and Ephesus, Turkey.[2] Sea-level rise is also an existential threat to low-lying nation-states such as Vanuatu and coastal cities including Alexandria, Miami, The Hague, Hong Kong, Rio de Janeiro, and Venice (the one in California *and* the one in Italy). And salt water wreaks havoc outside of cities, too, when it infiltrates farmland and percolates up through the ground into drinking-water supplies. Some

Nantong

Greater Shanghai urban area
below sea level in 2060
rural areas

Changshu

Yangtze River

Chongming
Island

Kunshan

Suzhou

Central Shanghai
and Pudong

Hengsh
Islar

Nanhui
New City

Jiaxing

The Greater Shanghai region is a flood plain
that is home to 40 million people, with Central
Shanghai a sunken doughnut hole at its heart.
Some global cities, including Shanghai, Tokyo,
and Los Angeles, have dramatically slowed their
past subsidence rates, but the damage already
done leaves them highly vulnerable. Other
cities, mainly in Asia, continue to sink.

SHANGHAI, CHINA: SINK, SANK, SUNK

The bottom line of keeping the environment safe is that the sword is high and heavy and heavy.

—Chen Jining, China's Minister of Environmental Protection, 2016[1]

In 2012, halfway through construction of Shanghai Tower, cracks began appearing in the surrounding pavement. This structure, the tallest pinnacle in the city's glittering skyline, lies at the heart of Shanghai's Pudong financial district, which over the past twenty-five years has grown into a dense mass of high-rises across the Huangpu River from the old city. A former marsh, Pudong is now home to more than three thousand buildings over eighteen stories high, all of which compress the silty soil beneath and cause the ground to deform.[2] But financiers are impatient and cracks can be covered, so Pudong's roads were patched and Shanghai Tower was completed, just a little behind schedule.

Once a walled garrison and provincial trading port, Shanghai grew in the late nineteenth and early twentieth centuries into a colonial mercantile city of five million and a major manufacturing hub. Such rapid, unregulated growth soon meant Shanghai was tapping out its freshwater aquifer, causing extreme land subsidence and by the 1950s, the city was sinking ten centimeters (four inches) per year. Alarmed, the government banned private wells in the 1960s and began requiring industries to inject their treated wastewater back into the ground, a procedure called artificial recharge. This dramatically slowed the pace of ground subsidence but did not stop it entirely. Looking back, researchers have deduced that two-thirds of the

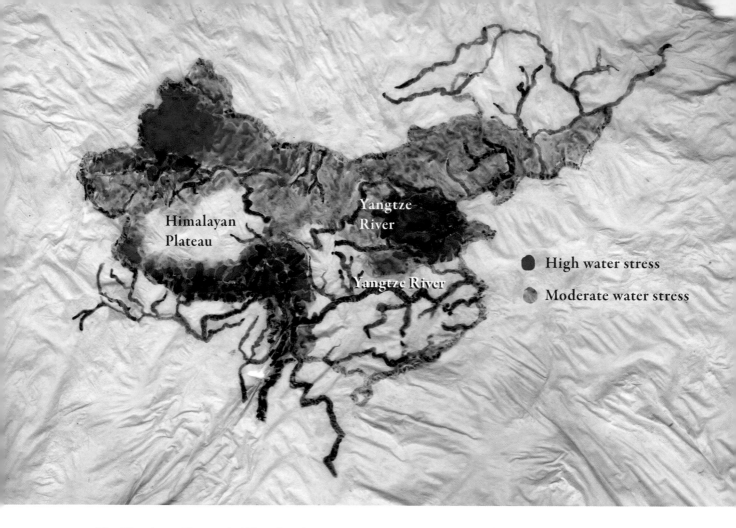

Himalayan
Plateau

Yangtze
River

Yangtze River

● High water stress

◐ Moderate water stress

The Himalayan Plateau in Tibet feeds sixteen major rivers in Asia and 1.7 billion people. As the glaciers melt, water security in China is at risk.

city's sinking has been caused by industrial groundwater extraction and one-third by the sheer weight of the built environment, which has only picked up speed in recent decades. Over the past eighty years, central Shanghai has sunk two to three meters (seven to ten feet) below its alluvial level at the confluence of the Huangpu and Yangtze Rivers. Scientists are warning that in the future "flooding will be severe."[3]

Living tissue, if it gets a little dry, can often rehydrate and recover if treated in a timely way. Dehydrated soils can sometimes

be minimally plumped by infusing water back into the ground. But the worse the depletion, the more misshapen and damaged the aquifer becomes, so most subsidence can only be stabilized, not reversed. Due to uneven subsidence and gravity's pull, Shanghai's aquifers now flow west rather than east toward the ocean as they used to, increasing the risks posed by saltwater intrusion from the rising East China Sea into the freshwater supply.[4]

Central Shanghai is now concave, shaped like a shallow rice bowl cover, according to

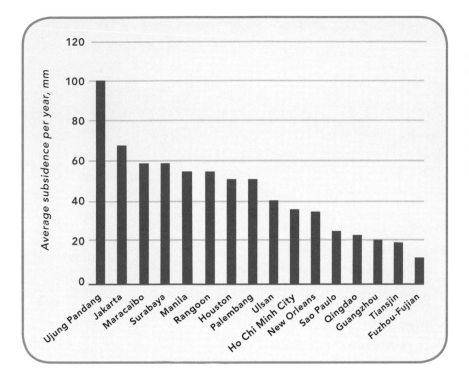

Annual subsidence rates in major port cities. By 2050, the combination of rising seas and sinking land will cost large coastal cities up to $1 trillion a year.

Tang Yiqun, a professor of geology at Tongji University.[5] Because of this, Shanghai is in danger of flooding from both rivers and the sea. The government has built an extensive and costly network of walled canals, dikes, tunnels, and dams to protect the city from catastrophe: flood walls along the Huangpu River are already seven meters (twenty-three feet) in places, some attractively disguised as shop-lined promenades. Overall, however, half the city's levees are only high enough to protect against a fifty-year flood, and the river is already higher than the surrounding streets at high tide. Two hundred kilometers of new seawalls are planned in order to defend growing exurban districts. Despite all this, a combination of heavy rains and high tides would leave the water with no place to go but into the low-lying central city. According to the Coastal City Flood Vulnerability Index, which evaluates many natural, political, and socioeconomic factors to assess systemic flood risk, Shanghai is now the most vulnerable city in the world.[6]

The entire Yangtze Delta Region of 150 million people lies, at most, just three to four meters (ten to thirteen feet) above sea level, with lower cities like Yancheng ("Salt City") climbing gradually toward ancient Suzhou and bustling Wuxi. The government's near-term plan is to foster the development of nineteen urban clusters, including Shanghai and its many surrounding cities, that will account for 90 percent of national economic activity.[7] Shanghai is predicted to double in size by 2050, just when scientists predict sea levels will begin rising quickly, making urban systems ever more vulnerable.

Stormwater levels show us the future of such places, when sea level will be the permanent challenge that occasional flooding presents now. Sea walls and dikes are near-term fixes, but they cannot contend with an ocean that is two meters (seven feet) higher, as is predicted in a hundred years or so, according to the Intergovernmental Panel on Climate Change's "business as usual" scenario. Even using the IPCC's most optimistic projections, in which humanity has already ceased all carbon emissions, sea levels will continue to rise until the year 2500.[8] Urban delta cities like Shanghai are the places where the convergence of sinking land, rising seas, increasing storm strength, and population growth come together as an unsolvable puzzle.

In 2005, President Hu Jintao visited Chongming Island, a rural Shanghai precinct, to announce construction of China's first purpose-built "eco-city," which he envisioned as a living example of *shengtai wenming*, or ecologically harmonious socialist society.[9] He described ecological civilization as "a new choice for human society after agricultural civilization and industrial civilization"—a way to bring together economic strength and environmental good.[10] President Xi Jinping, Hu's successor, reinforced this ideal in 2018, when he stated that "clear waters and green mountains are as valuable as mountains of gold and silver."[11] Interpreters of Xi Thought have taken this to mean that environmentally sustainable projects can also be profitable projects. Xi led a major governmental reorganization in 2018 intended to speed up reforms to environmental law and enforcement, taking up the mantle of international leadership the United States has set down.[12]

Development continues at a dizzying pace in Shanghai, with emphasis on finance and tech rather than on water-consuming industries like manufacturing. Major construction sites attract millions of "floating people," who move from project to project, building residential areas that extend like tentacles through the greater Shanghai plain.[13] The city has drained and filled the wetlands at its eastern edge, building Nanhui New City on the shore of the East China Sea, and it is expanding nearby Hengsha Island, another sandbank in the Yangtze, for urban expansion. For now, the Communist Party is banking on the least ecologically sustainable solutions to keep its civilization safe from coming storms and sea-level rise.

~~~~~~~~~~

Key Term: Ecology

The Greek roots of *ecology* translate as "the study of the household or home." Ecology, then, is the study of our planetary home with its many rooms, inhabitants, and relationships. British ecologist A.G. Tansley proposed the word *ecosystem* in 1935 to acknowledge the complex relationships among living and non-living matter, and American ecologist Raymond L. Lindeman extended Tansley's thinking by describing an ecosystem as the transformation of energy itself through several "trophic levels," or layers of the food web.[14]

For decades, economists, ecologists, and other professionals have attempted to quantify the utility of ecosystems to people as the value of "ecosystem services" and "natural capital." The U.S. Climate Resilience Toolkit

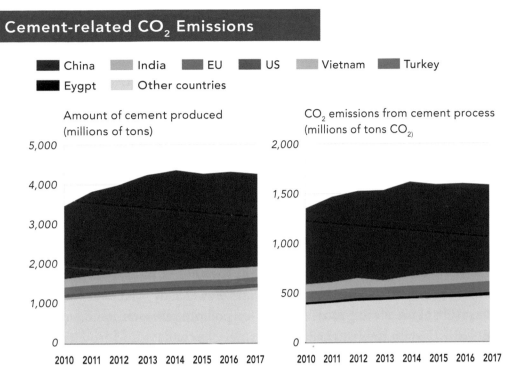

Cement-related CO$_2$ Emissions

Legend: China, India, EU, US, Vietnam, Turkey, Eygpt, Other countries

Amount of cement produced
(millions of tons)

CO$_2$ emissions from cement process
(millions of tons CO$_2$)

Cement production makes up 8 percent of global carbon emissions.

currently estimates the monetary value of the world's ecosystem at $124 trillion to $145 trillion per year—a shockingly low number in the scheme of the global economy.[15] These numbers reduce nature to its parts, in transactional service to us, and fail to consider factors like overall ecosystem health (which benefits us) and the needs of other living things (which also benefit us). If systems are by definition more complex than their components, such calculations necessarily undervalue the world.

Other ways of valuing ecosystems exist. "Deep ecology," an approach conceived by Norwegian philosopher Arne Naess, understands the earth as an interdependent system of living things that all have inherent, though not equal, rights.[16] Though

philosophy and economics may seem distant and distinct fields, the former always underlies the latter, especially as they relate in the field of ecology.

A View from 2050

Though Shanghai's authorities woke to the damage being done by groundwater extraction in the 1960s, they only got serious about addressing it in 2005, doing so with a combination of regulations and fines that reduced subsidence to "a safe level," according to party officials. But it took the imprisonment of several high-profile business leaders who flouted these

HAMPTON ROADS, VIRGINIA: BYE, BYE, BIRDIES

> Someone once wondered why it is that if a work of man is destroyed, it is called vandalism, but if a work of nature . . . is destroyed it is so often called progress.
>
> —Jane Goodall, *Seeds of Hope*[1]

Hampton Roads, Virginia, is home to about 1.7 million people, and that number is growing. The wetlands that surround Hampton Roads are home to about one million birds, and that number is shrinking.[2] Here in the Chesapeake Bay region—which contains the largest estuary in the United States—new housing developments and rising tides are squeezing the wetlands and their inhabitants from all sides. As the landscape changes, the hundreds of species of birds, fish, amphibians, mammals, reptiles, insects, and plants that live here must either adapt to new conditions in their current home, relocate to a new one, or die. For many of these species, relocating is geographically impossible and

biological adaptation takes too long. That leaves only dying.

Estuarine wetlands like those of the Chesapeake are ruled by the tides. The saltiest, wettest, and lowest-elevation zone is the mudflat, where wading birds like herons and egrets feast on the critters hiding in the mud that is exposed at low tide. The middle of the marsh spends part of each day under water, and salt-tolerant plants like cordgrass and pickleweed function as nest and nursery for fish, mammals, and birds. The highest areas of the marsh are inundated by salt water only a few times each month during the highest tides, and also during storms. The salinity of the water in an estuary exists on a gradient as well, from ocean-level salty

Potomac
River

Maryland

Rappahannock
River

Chesapeake Bay

Virginia

Delmarva
Peninsula

York
River

James
River

Atlantic Ocean

Hampton
Roads

Wetlands

Urban areas

Norfolk

Virginia
Beach

Chesapeake

Half of the world's 820 million hungry people live in Asia, where rice is the staple, and sometimes only, crop. In the Mekong Delta, freshwater flooding meets saltwater intrusion, putting basic food security at risk for millions of farming families. Looking at global crop trends, Cynthia Rosenzweig, a senior researcher at NASA, says, "The potential risk of multi-breadbasket failure is increasing."

Upper Mekong River

Phnom Penh

Cambodia

Vietnam

Ho Chi Minh City

Mekong Delta

Cần Thơ

Bến Tre

The main crop in painted areas is rice.

High risk of salinization

Slight risk of salinization

Risk of freshwater flood > 1m

Risk of freshwater flood > .4m

Annual flood line in 2040

National border

BEN TRE, VIETNAM: DOING MORE WITH LESS

In normal times, the scholars rank first, the farmers second.
But during a famine, farmers are first, scholars second.

—Vietnamese proverb[1]

The day starts before dawn, when Nguyen Duc rises to make tea and check on his garden plot, weeding and watering here and there, before making a morning offering of fruit and flowers to his ancestors at the family altar. For the past week, he and his wife, Anh, have been plowing their rice paddies. It's planting season, and everyone is needed in the fields, here at the eastern edge of the Mekong Delta.[2]

Duc puts on his conical *don la* hat to protect himself from rain and sun and walks with his wife and sons along the low dikes to their paddy, past coconut trees and wandering ducks. Shedding his flip-flops and grabbing an armload of bundled rice seedlings, Duc steps into the knee-deep mud, scattering the small green tufts across the sticky surface of the long, narrow plot. Holding the first bundle in his left hand, with his right he plunges seedling after seedling into the silt, six inches apart, two per second, almost faster than the eye can see, bundle after bundle. Anh and the children weave back and forth across the paddy too, all morning long, until the heat of day signals it is time to harvest and prepare their midday meal of spicy sautéed vegetables and herbs, rice, and a steamed fish, fresh from the river. The afternoon follows the morning: planting, watering, weeding, fertilizing, tending. Soon the rains will come, flooding the fields and feeding the garden.

Half as productive   Twice as productive

Projected agricultural productivity in 2050. As weather patterns change, some regions will become more productive, but many of the most populous places will face droughts and hardship.

During the dry season, Duc cuts the thick brown mounds of rice near the ground, so a second crop can regrow from the established roots. When he was a child, he and his father only grew one rice crop a year. But after the American War, when bombs rained from the sky and the land was poisoned by Agent Orange, the Communist Party had taught farmers to grow two, providing fertilizers and insecticides to increase yields. It had taken decades for the land and its people to recover, but there had been a few good farming years. Hunger was rare by the time Duc married Anh, who grew up in the next village. And when the government introduced *doi moi*, the 1986 policy that allowed farmers to sell their crops and keep the profits, life became easier yet again. Eventually, their family was able to save enough to buy their buffalo.

But now the province is once again being contaminated, this time by changing weather and salty seas. In 2013 the dry season came a month early. Higher air and water temperatures led to more evaporation, and the resulting lower river levels drew more seawater farther upriver. Then the wet season started late, which delayed the dilution of the salty soil that was necessary before planting could begin. Everyone had to take out loans to buy the food they couldn't grow. In 2016, the drought parched the land worse than before, chemically "burning" the fields of rice overnight when salt water intruded into the paddies. "The water is salty every year, but it's been worse in the last three years," said Hai Thach, a Mekong Delta farmer. "I'm scared because I cannot live without rice. There's usually enough fresh water to use, but the rainy season has come late. The rain has not diluted the salty water enough to grow rice." Many farmers lost their entire crop in this period, including Duc. To avoid having to sell his buffalo, he moved to the nearest city, Cần Thơ, and took a temporary factory job. But his ties to land, family, and ancestors are stronger than his fear and uncertainty. Duc may spend a season or two in the city, but he will never leave Bến Tre altogether.[3]

Both drought and flood are more common in the Mekong Delta now; the 199-day-long wet season from late April to mid-November is six days longer than it used to be.[4] Duc's rice yields have dropped nearly by half over the past five years, as weather has grown unpredictable and salt water has reached farther inland.[5] Some of Duc's neighbors—those who have savings and the gumption—are investing in tiger shrimp farms, which requires paying to dredge out their paddies, plus the costs of new equipment and special feed. The entire southern perimeter of Vietnam is rimmed with thin dikes that barely distinguish floodable shrimp ponds from the sea. But this is not a long-term solution. As one farmer reported, "We had to sell our fishing boat to pay to dig the [shrimp] cultivation pool and also had to pay someone to teach me how to do it. . . . It was expensive, and I had to get the shrimp food and medicine on credit. But last October, the sea washed out all of our shrimp. We lost them all. We saw the water rising up and getting closer and closer, but we couldn't do anything about it. This season, we have been forced to just dump the shrimp in and let them grow with no fans, medicine, or special food."[6] Though able to make more money with shrimp in a good year, Delta farmers are less food secure than they were as rice cultivators.

Recently, the district's rural development bureau has been encouraging farmers to diversify their crops to include coconut, grapefruit, and grass for grazing buffalo. The People's Committee of Bến Tre reports that 7,500 hectares of rice were converted to other crops between 2015 and 2017, which has increased overall agricultural yields in the province by 63 percent.[7] Duc thinks he might try planting some fields with grass next year, which will spread his risk without large upfront costs.

A one-meter rise in sea level will force around 30 percent of local residents to migrate, according to the Mekong Delta Commission, an intergovernmental body.[8] Tran Thuc, the director general of the Vietnam Institute of Meteorology, Hydrology, and Environment, cautions, "If there was a one-meter [sea-level] rise, we estimate 40% of the delta will be submerged. There is also the threat of cyclones and storms linked to climate change. The people in this area are not prepared for any of this." The central government estimates that sea level could rise by 75 to 100 centimeters (2.5 to 3.3 feet) by the end of the century.[9] Yet "these predictions are quite conservative and moderate," says Kien Tra-Mai, a climate specialist with the Mekong River Commission. "Other forecasts afford much worse scenarios."[10]

Duc and his family are among the eighteen million people who live in these lowlands. They produce half of Vietnam's rice and two-thirds of its fish and shrimp on only 12 percent of the country's land.[11] Yet over the last decade, more than twice as many people have left the delta as have arrived, citing climate change as a primary reason for migration.[12] Most rural families move to nearby urban slums. Exacerbating local flooding problems, in some parts of the Lower Mekong estuary, subsidence caused by groundwater pumping is making the land sink up to ten times faster than the sea is rising.[13] For subsistence farmers in coastal food-basket regions, tenuous livelihoods are becoming more so.

~~~~~~~~~~

Key Term: Retreat

Retreat is both verb and noun, the act of withdrawing and a place of quiet reflection. In its military connotation, retreat is often associated with defeat. It is a term avoided by many American politicians and planners, where words like *de-densification* are used to

describe the act of moving people away from the most flood-prone areas. In the United Kingdom, *managed realignment* is the palatable term because it implies a well-ordered process that will allow the public to keep calm and carry on. In developing countries, where most people have little money and no power, semantics mean far less. Over the next several decades, tens of millions of rural people in much of the coastal world will be overwhelmed by their circumstances and displaced by either disaster or poverty as seas rise and storms surge.

Whatever the preferred term, some communities are beginning to consider the idea of voluntary retreat as their shoreline homes are more frequently flooded. Some come together to seek government buyouts and relocation assistance because there is more power in strategy and forethought than in crisis response.[14] Far from failure or defeat, retreat can thus take on its other meaning: moving to safer places on higher ground in which to build healthy communities. A million tiny retreats of this kind can and must reshape the world's coastline communities, moving us away from danger zones mired in the twentieth century's appetites toward a more functional and sustainable society.[15]

~~~~~~~~~~

A View from 2050

In the early twenty-first century, Vietnam's Communist Party invested millions to improve roads and dikes in the Mekong Delta. But five thousand miles of dikes was

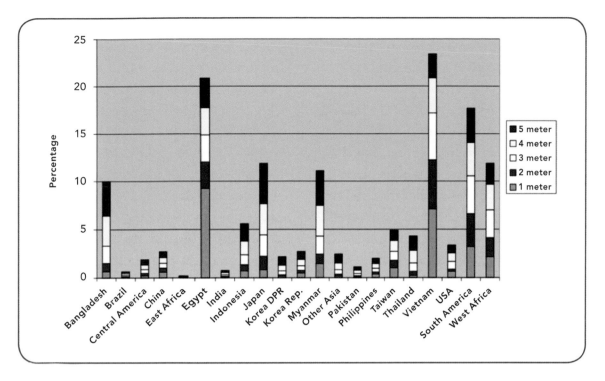

The percentage impacts of sea-level rise on agricultural land.

a tall order, so when the government was approached by the Netherlands in 2010 with a proposal to develop a Mekong Delta Plan that might yield international funding, the party agreed.[16] The first Dutch attempt applied homegrown water management priorities and disregarded the political realities in Vietnam, so the project stalled and nearly fizzled out. But two Dutch consultants revised the plan unbidden, and then brought together a strategic group of academics and officials—the "retired reformers"—to put in a good word with party leaders. The consultants also revised the Dutch approach, which featured restored wetlands and other luxuries, to focus instead on Vietnam's main priority, socioeconomic development.[17] Their report proved influential, but the Vietnamese

government, sensitive to colonial legacies, did not accept it wholesale. Instead, they used it as a starting point for their own plan, which further emphasized economic growth.

In 2017, Vietnam's government issued Resolution 120, which proposed a vast network of economic corridors across the delta that would move the region toward light industry and food processing for export and boost GDP while reducing disaster risk.[18] It ordered that officials "select models of nature-based adaptation, environmentally sound and sustainable development, on the basis of actively living with flood, brackish and salt water," including worst-case scenarios.

But the balance of economic development and climate adaptation proved elusive in the 2020s and 2030s because central planners

Ducks swim in brackish, algae-rich shrimp ponds, fertilizing the rice crop, a farming technique that yields several sources of income.

didn't consult local leaders about the truth on the ground, and corrupt politicians drew lines on maps that benefitted their business interests and friends. Storms lashed the region and salt water infiltrated the rice paddies. Light industry could not compensate for the lost lives and incomes in the outer delta. In a poor country like Vietnam, which was striving to catch up with its Asian Tiger neighbors, how leaders addressed the climate crisis in those decades influenced almost every metric of national growth and success thereafter.

Rice farming continued to be a dirty business. Farmers either burned their fields at the end of each crop cycle or allowed them to rot, releasing methane, so the rice crop continued to account for one-quarter of the country's greenhouse gas emissions.[19] Heavy reliance on fertilizers and pesticides meant that most farmers refused to eat what they grew. In Bến Tre, the party's district leader was keen to try out some cutting-edge farming techniques

to save his district from depopulation. He implemented several "farmer field schools," where agricultural extension teachers demonstrated integrated pest-management techniques and best planting practices. Duc's teacher distributed test kits that showed if the river was too salty before opening his irrigation gates. Duc's farm was also selected by the party for the innovative rice-duck-shrimp program initially piloted by Seed to Table, a Japanese nonprofit.[20] He planted a more salt-tolerant variety of rice, while growing shrimp in the same warm, brackish, flooded field and also herding a team of ducks to and from the paddies each day to eat the surface algae and fertilize the rice with their droppings. Duc had been skeptical, but to his surprise the system worked well, used far fewer chemicals, and was more profitable for more of the year. His grandchildren were able to complete high school and then technical college.

By the late 2030s, Duc, in his seventies,

was so frail that he could no longer plant or harvest, but he still tended the family's vegetable garden. He had seen so much in his life: the war, the communist collectives, the Green Revolution, the capitalist reforms, new farming techniques, the big storms, and the coming of the waters. That last had been so difficult, so unpredictable, and had caused so many people to move away. His family had been very fortunate, but he saw, as he neared the end of his life, that it would not last. Though he never said so to his grandchildren, he knew that he would be the last in his family to live, farm, and die in Bến Tre.

THE THAMES ESTUARY, BRITAIN: FROM GRAVESEND TO ALLHALLOWS

Here is what I fear: other people's cowardice.

—George Monbiot, British activist and writer[1]

On June 25, 2019, six-year-old Leo Mansten stood before the Gravesham Borough Council, one of the local governments on the eastern edge of London, to ask a question. "The planet is getting hotter," he said. "People are dying. Would you consider having a selection of children and young people on the council to help make decisions about climate change? It will affect us the most."[2] Leo's mother, Laura, had not been particularly active in local politics until she attended the million-strong Extinction Rebellion march in London in April 2019. She was so disturbed to learn how urgent the situation is that she decided to get involved in the move to get local government councils to declare a climate emergency, and Leo

had wanted to come along. That night, the Gravesham Council demurred on youth enfranchisement but voted in favor of the declaration, as Parliament had done unanimously a few months before.[3] Yet the gap between such easy gestures and the U.K. government's actual policies and spending priorities remains as wide as the expansive tidal sands of the outer Thames River estuary.

The Thames winds its way from Gloucestershire toward the English Channel, its last 104 kilometers (65 miles) running through central and east London, and then past industrial and rural landscapes, to the sea. Protecting central London from storm surge and sea-level rise is the Thames Barrier, which was completed in 1983 in response to

London has grown up along the Thames River for more than two thousand years. Constraining the river with dikes and embankments helped protect people from floods for centuries. But as seas rise, the Thames Barrier must be closed more often, and floods will become far more frequent and severe for those living near the river.

Maldon

Greater London area

Southend-on-Sea

Shoeburyness

Newham

Thames Estuary

Thames Barrier

Allhallows

Sheerness

Gravesend

Rochester and Strood

- Urban area
- Land predicted to experience floods every decade by 2060
- Rural areas

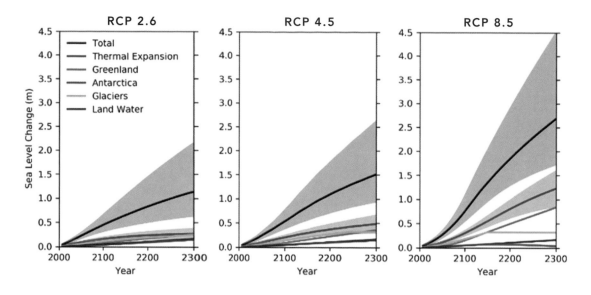

Projected global mean sea-level change to 2300. Most scientists consider RCP 4.5 likely, whereas RCP 2.6 is highly unlikely, and RCP 8.5 is cited in British planning documents as "business as usual."

the deadly flood of 1953 and lies just east of the City of London financial district. Tides in the Thames can rise and fall seven meters (twenty-three feet) in a day, with low tides exposing stony beaches and high tides lapping at the top of London's river walls. Over nearly forty years, the barrier has been closed 193 times (as of September 2020), saving London from many catastrophic floods. [4]

East of the barrier, however, there are few protections from the sea, where more than 1.5 million people live on land lined by seawalls and dikes. The boroughs of Newham and Greenwich, which lie on either side of the barrier, are at greatest risk of severe inundation; parts of London's Royal Docks neighborhood will be below the tide line when seas rise one meter. This inner estuary includes many of London's edgelands, places where London has sent its sewage and garbage for

centuries, creating a topography of low, rolling hills where there were once only marshes. The Barking sewage works lies adjacent to the River Roding, a Thames tributary, and just upstream from the Rainham Marshes and Landfill. Across the river, Thamesmead, a town built on infill, is already two to three meters below sea level at the highest tides. Here flood depths in a surge could exceed five meters. [5]

The outer estuary, though it is just a few kilometers farther downstream, is a landscape Charles Dickens called "the dark flat wilderness . . . intersected with dikes and mounds and gates," the Thames a "low leaden line beyond." [6] This foggy expanse of tidal mudflats and shifting sands is interrupted only by the odd cargo ship, oil terminal, and gas storage facility. Few towns dot the shore, except at higher promontories like Gravesend and poorly sited holiday villages like Allhallows-on-Sea. Though the government no longer parks prison ships in the mud for men to languish and die in, as in Victorian times, as oceans rise, these current and former marshlands could one day regain their earlier reputation as breeding grounds for waterborne disease, including cholera, typhoid, and malaria, which Dickens refers to as "the ague."

The British love a good plan, so in 2012 the government's Environment Agency published *Thames Estuary 2100*, which provides guidance to other agencies and local governments on regional priorities. The plan acknowledges that twenty to ninety centimeters (eight to thirty-five inches) of sea-level rise is all but inevitable by the end of this century.[7] It focuses on developing the estuary by converting its rundown communities, low-value lands, and industrial estates into high-density hubs of commuter commerce. But it does not require this new development to be on higher ground or even slightly removed from the river's edge. Brand-new malls and luxury high-rises already abut the river in towns like Erith and Greenhithe. Yet despite its important implications, *TE 2100* received little attention upon its release and remains all but unknown today outside planning circles. Though it primarily assesses flood risk, this report also creates a de facto plan for sea-level rise management in the region—which is to hold the current shoreline against the rising tide, whatever it takes.

TE 2100 was reaffirmed as written in 2016, despite rapidly improving science and worsening sea-level rise projections. Its restated goal is to maintain and improve the existing national system of dikes, flood gates, and barriers "as the optimum approach for the first 60 years of our Plan, with new arrangements required by 2070." The only evaluation process is a decadal review of the science and a grand review of the plan in 2050.[8] The government is not requiring more frequent updates or opportunities for public scrutiny because so few citizens have demanded it.

Developers have taken note, though. The updated London Plan, being finalized in 2020, is rife with mentions of "opportunity areas" east of the city, but it relies on the *TE 2100* data to "ensure that London is protected until the end of the century."[9] Another report, this from the Thames Estuary Growth Commission, envisions filling in all the estuary's existing gaps with roads and bridges, flats and shops, businesses, industries, and universities to orbit around London, resulting in 1.3 million new jobs and 4.3 million people by 2035, triple today's population. No one is talking about where and how all of those jobs and people will be moved once the water comes.[10]

The difficult truth is that not every shoreline community will come out on the winning end of the U.K. government's cost-benefit analysis, and many places will have to be abandoned.[11] Those that are chosen for defense will become fortress-cities that sop up most of the budget in their struggle for survival, while others will languish, underserved. A different approach would acknowledge that sea levels around Britain will continue to rise for centuries and will erase many of today's hard lines between land and sea. It would return large areas to the tide

Thames River seals have begun repopulating the outer Thames Estuary in recent decades.

and create or reestablish soft infrastructure like marshes, sand barrier islands, beaches, and mudflats in as many places as possible. Instead of inviting millions into the path of the water, it would prepare for the gradual, rational relocation of millions to higher ground.

~~~~~~

Key Term: Transilience

Literally meaning "to jump ahead," *transilience* is an old term from geology that refers to the abrupt transition from one layer of rock to another, indicating a moment of dramatic change.[12] It is ever more clear that humans are indeed a geological force, changing the climate, weather, and geography in ways that will mark the fossil record forever. Therefore, *transilience* is a word worth redefining for our time. In addition to being *resilient*—which emphasizes returning former levels of function or making incremental adaptive changes—we can also take

the leap toward many new (and some very old) approaches that can become the basis for surviving the climate crisis and thriving thereafter.

For instance, if we only rely on sea walls to protect us, we will inevitably lose, because water always finds its level. We will spend most of our money and time, and potentially many lives, being overwhelmed by a body infinitely more powerful than our own. But if we imagine, design, and build places and communities that are synchronized to live well with the ocean—and if we take only what is needed rather than everything possible—we can emerge with the knowledge, wisdom, skills, and purpose we require to navigate the storms that lie ahead.

~~~~~~

A View from 2050

In the United Nations' mid-century review of how nations had handled climate change, Britain ranked last on the list of adaptive capacity in developed countries. Part of this was due to its island geography, which had more shoreline per capita than most of its peers. But the majority was the result of political and planning choices the British government had made decades before, in the 2010s and 2020s.

As predicted, the years immediately before and after Brexit and the COVID-19 Depression sucked all the energy out of the country, leaving little space for thoughtful climate adaptation planning or serious infrastructure investment. The 2018 report from the Committee on Climate Change (CCC), an independent advisory panel to the government, was the only candid critique in this period, outlining what was at risk: up to 1.5 million properties (including 1.2 million homes) in the flood zone and over 100,000 properties at risk from coastal erosion, as well as 1,600 kilometers of major roads, 650 kilometers of railway line, ninety-two railway stations, and fifty-five historic landfill sites. The report noted that the government's priority of reaching housing targets and economic goals was in direct conflict with sea-level rise planning. It also mentioned that few people had any insurance to cope with losses, as in most of Europe.[13]

Throughout the 2020s, significant gaps remained in the national climate strategy, and the idea of building a wall around a large portion of the island nation persisted.[14] Thousands of Extinction Rebellion activists continued gluing themselves to roads and buildings and blockading oil tankers from entering British ports. They always made the same three demands: that the government tell the truth about the climate crisis, that it work urgently toward decarbonization

by 2025, and that a Citizens' Assembly be formed to bring people's voices to the heart of government climate policy. Precious little progress was made, followed by the implosion of Britain's gangrenous Conservative and Labour parties, which created a political vacuum that lasted years. National unity governments formed and collapsed, and the climate crisis didn't make many of the headlines as the sea continued to slowly rise. Half a centimeter a year didn't seem like much, but the number of minor floods increased steadily, putting town folk especially on edge in the winter.[15]

On the upside, a retrenched economy made it relatively easy for the United Kingdom to meet its carbon-reduction mileposts in the 2050 Paris Accord. The government crowed about its carbon-reduction accomplishments, yet fell further behind in infrastructure investments. As the people became poorer, many returned to traditional pastimes: allotment gardening, camping holidays in Wales, hunting and fishing, preserving food. Something in the British character liked mustering for a challenge and defying poor odds, uniting the politically divided country.

The event that finally put sea-level rise in the headlines was the reappearance of malaria in the Thames Estuary in 2035 after 150 years, brought by *Aedes albopictus*, a mosquito that had been marching farther north each year. Even experts at the Hospital for Tropical Diseases were caught by surprise when thousands of people were admitted to County Kent's emergency rooms with high fevers, rashes, headaches, vomiting, and severe joint pain. A long hot summer had

Jakarta, Indonesia

Bangkok, Thailand

● Ocean ● SRTM projections   ● CoastalDEM projections ☐ Land

Pearl River Delta, China

Sundarban Bangladesh

In recent decades, Shuttle Radar Topography (SRTM) data greatly underestimated sea-level rise in coastal megacities because building roofs read as "land" from space. New Digital Elevation Model (DEM) tools allow scientists to map topography with greater accuracy, showing far greater risk that previously understood, seen here in red.

followed the stormy winter, leaving a lot of standing water in the outer estuary, a perfect environment for the Asian tiger mosquito. Doctors also identified both chikungunya and dengue, two of the twenty diseases carried by *Aedes albopictus*. While many people were treated and sent home, seven hundred children died.[16]

Heartbroken and outraged by the Lost Children of Kent, Leo Mansten—whose childhood had been marked by the COVID-19 Crisis of the early 2020s—returned home after graduating from university in order to stand in the 2036 general election. As the Green Party candidate for Rochester and Strood, he won handily, along with many other Greens and Liberals that year, who represented voters' desire for an ambitious overhaul of national politics and policy.

The new Green-Liberal coalition government undertook a complete revision of the United Kingdom's sea-level rise strategy. Fueled by a new generation of Members of Parliament and civil servants, *Plan 2080* had teeth. Compulsory public purchase of lands below the +2 meter flood line (over 2000 levels) made a national-scale "managed realignment" possible, and over a decade it gave much of the Thames Estuary back to the tide. Above the five-meter mark, the government bought land for new towns with good public transit and access to green space, using many of the old Roman roads from London as its wheel and spokes.

By 2040, the scientific consensus projected that nearly two meters of sea-level rise was likely by the end of the century. The Green Party used this expert consensus to withstand pressure from more conservative colleagues to build a massive sea barrier at the mouth of the Thames between Sheerness and Shoeburyness, which would have turned the river into a stagnant, smelly bathtub for parts of each year. Leo, who became minister of the environment in 2044, persuasively explained to the public that, with seas set to rise for hundreds of years yet, the wise choice was to work in partnership with nature, not against it. The government did, however, build a new emergency barrier at Long Reach in the inner estuary, since the old barrier was well past its 2030 sell-by date. Should the new structure fail, central London would be under six feet of water within an hour, but even the Greens could not yet face the fact that most of London's major landmarks faced a very uncertain future.

In last spring's power tussle, Leo stood to become the leader of the Greens, and thus prime minister. In his victory speech he quoted the French thinker Bruno Latour: "The time is past for hoping to 'get through it.' We're going to have to get used to it. *It's definitive.* The imperative confronting us, therefore, is to discover a *course of treatment* —but without the illusion that a cure will come quickly."[17] Leo went on to speak hard truths about the sacrifices that would be needed to pull through the climate crisis, pulling no punches but inspiring a nation to seek solidarity and courage during the difficult decades ahead.

Nagoya

Ise Bay

As a UNESCO World Heritage Site, the Grand Shrine at Ise is meant to be protected in perpetuity. Climate scientist Ben Marzeion asked what could happen to these forever sites over the next two millennia and found that, even if we stabilize temperatures at present levels, around forty sites will slip under the rising tide. With a rise of 3C°, 136 sites drown, and three dozen countries lose 10 to 50 percent of their land.

5 meters sea-level rise

Shinto shrines

Ise town

Geku Shrine

Grand Shrine (Ise Jingu)

Pacific Ocean

ISE, JAPAN: TRADITION FOR THE FUTURE

Every major religion today emerged out of a previous ecological and social collapse. We've been through this before. . . . Out of these disasters the faiths had to reconstruct meaning, had to reconstruct a story of who we are and what our relationship is with the rest of life on earth.

—Martin Palmer, Alliance of Religions and Conservation[1]

At the turn of the millennium, leaders of Japan's indigenous Shinto religion attended a multi-faith environmental conference in Kathmandu, where "they had a revolutionary epiphany," according to journalist Paul Vallely.[2] They declared that spirits and deities did not reside just in the natural phenomena of Japan, but all over the world. This announcement marked the opening of their ancient, introverted tradition to the global interfaith movement for the first time. Then, in 2014, Jinja Honcho, the Association of Shinto Shrines, offered to host a gathering of global religious leaders at the Grand Shrine, Ise-Jingu, the most sacred site in Japan. The

topic: the role their diverse religious institutions could play in advancing environmental stewardship worldwide.

To open the event, dozens of priests, swamis, ministers, imams, and monks in their ceremonial saris, robes, cassocks, and tunics walked across the sacred Uji Bridge. The bridge, along with all 135 of Ise-Jingu's shrine buildings, had just been freshly reconstructed in exact replica out of cypress logs using ancient techniques, as they have been every twenty years since the original shrine was built in 690 C.E.[3] After crossing the holy Izuzu River, the leaders were met by a Shinto priest who bowed deeply, twice, and allowed them to enter the shrine, which had until that day been strictly closed to outsiders for more than two thousand years.[4] With this simple gesture, Shinto joined the global interfaith environmental movement, which has been growing quietly around the world for decades.[5]

Like many other religious and cultural landmarks, the Grand Shrine lies all too near the sea. The Main Shrine, Naiku, is a little bit up the hill from the coast, some twelve meters above Ise, the sleepy, low-rise town of 125,000 that grew up around the pilgrimage route. The second-most important site in Shintoism, the Outer Shrine, or Geku, lies at the edge of town, just three meters above sea level. Geku houses Toyo'uke, the goddess of food, cloth, and shelter, who is Amaterasu's (the goddess of the sun's) close companion. Every day, priests in starched white robes offer rice, flowers, fish, and prayers at both shrines.

Most of Ise town is already below sea level, separated from the ocean by an eight-meter (twenty-five-foot) cement berm that was built to defend the town from typhoons like Vera, which killed five thousand people and injured forty thousand in 1959.[6] In a country frequently visited by natural disasters, including earthquakes, tsunamis, and cyclones, it is understandable that people have long sought to appease the gods. But Japan's usual dangers now include more insidious emergencies too. For every degree Celsius global temperatures rise, scientists calculate that sea levels may, over time, rise by 2.3 meters (7.5 feet).[7] What becomes of holy land when it drowns?

In Shintoism, the gods, or *kami* (meaning "that which is hidden"), are simply higher manifestations of the sacred or mystical core present in everything. *Kami* are often embodied in specific rocks, trees, mountains, and rivers, and shrines are built at sites where a *kami* is especially present. Eighty-one thousand shrines commemorate the *kami* in Japan, creating rare green spaces in crowded cities and quiet places of contemplation in most villages.[8] But the *kami* are not supernatural, eternal, or all-powerful. They are vulnerable to injury, suffering, and death, and they can make mistakes and behave badly. Buddhist temples are often in close proximity to Shinto shrines, and over the centuries Buddhism has blended with Shinto, in part because both embody the Japanese understanding of impermanence and the interrelationship of each to all. Though few Japanese regard themselves as religious, most people visit shrines and temples regularly to make an offering or pray for health, prosperity, and happiness.

Yet Shinto's environmental awareness is not a strong ethos in Japanese culture. In 2015, to coincide with the Paris Climate talks, several environmental groups planned a large rally in Tokyo, expecting ten thousand people. Just one thousand came, including many expats from Europe and North America. "It was quite far from what we were trying to get," said Ayumi Fukakusa, an energy and climate campaigner with Friends of the Earth Japan.[9] Not many showed up to the Youth Climate Strike in Tokyo in September 2019 either, despite huge crowds elsewhere in the world.[10] According to a Pew Center poll, the elderly in Japan are more likely to be concerned about the climate than young adults, and levels of interest are falling, from 90 percent to 75 percent among eighteen- to

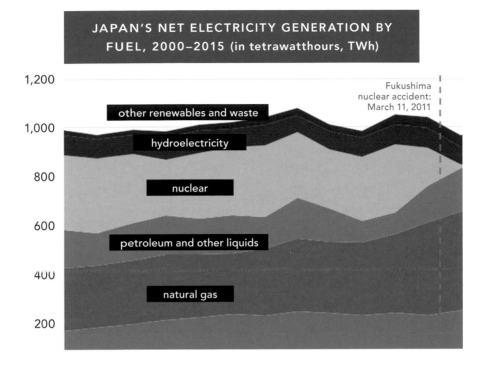

JAPAN'S NET ELECTRICITY GENERATION BY FUEL, 2000–2015 (in tetrawatthours, TWh)

Fukushima nuclear accident: March 11, 2011

other renewables and waste

hydroelectricity

nuclear

petroleum and other liquids

natural gas

1,200

1,000

800

600

400

200

Japan recently pledged to become carbon-neutral by 2050, largely by building more "clean coal" power plants. But this processed coal often emits more nitrous oxides than old, raw coal plants, and most fail to reduce emissions by even 20 percent. The government has no plausible strategy to reduce fossil fuel dependence.

twenty-nine-year-olds over recent years.[11] Perhaps this is because focus has moved to coal-fired energy production since the Fukushima nuclear disaster, or it is a result of civic reticence about marching in the streets. Or maybe the climate crisis does not conform to the usual Japanese emergency: no sirens have gone off to alert people of the incoming wave, no face masks are needed, the earth has not given way.

Key Term: Nature

The word *nature* shares a common root—*nasci*, to be born—with *nation*, *native*,

and *innate*: a quality of inherent, inclusive, foundational being. However, a longstanding, dualistic view of man versus nature, good versus evil, and subject versus object has set humanity against its own origins in the natural world, creating a tension that often leads us to seek mastery and possession. Even Romanticism, with its awe and reverence of nature, understands nature as separate from humanity, and thus only knowable as an Other. Many thinkers have questioned this stance toward nature as property or resource, seeing instead the scientific fact and philosophical value of entanglement, relationship, and belonging. French philosopher Michel Serres writes that nature is "caused, causing, all things in the world ensue from each other, chained together."[12] In *Laudato si'*, Pope Francis equates all marginalized bodies when he says, "Earth, burdened and laid waste, is among the most abandoned and maltreated of our poor."[13]And physicist David Bohm's theory of "implicate order," in which mind and matter manifest according to the same set of principles, leaves no space for nature as Other.[14] Starting with an interconnected, process-driven universe both more accurately describes the world seen by physicists and allows us to consider restoring both non-human agency and the sovereignty of nature

itself. Unravelling this complex concept to find our place of belonging within the natural world is an essential step in transitioning toward systemic, planetary health.

A View from 2050

Japan's early twenty-first-century climate leadership came from a surprising place: corporations. In 2019, the country hosted the Task Force on Climate-Related Financial Disclosures (TCFD), a dry but necessary

Many UNESCO World Heritage Sites are vulnerable to erosion and flooding caused by rising seas. In the Mediterranean alone, the list includes Carthage, Arles, St. Mark's Cathedral (and all of Venice), Pisa's Duomo and tower, Ravenna, and sites in Instanbul, Haifa, and Ephesus. Rapa Nui's monolithic statues are being eroded by higher waves, and the Elefanta Caves in India will soon be inundated.

Temperature rise predicted to impact UNESCO World Heritage Sites

Degrees C
- 0–.5°
- .5–1°
- 1–2°
- 2–3°
- 3–4°

gathering of executives to "develop voluntary, consistent, climate-related financial risk disclosures for use by companies in providing information to investors, lenders, insurers, and other stakeholders." In other words, it soothed C-suite jitters about what environmental details each company would need to share about their businesses in response to growing customer and investor demand for insight and oversight. In its early years, Japanese business leaders were the TCFD's largest block of supporters, with more than 200 signatories, compared to 107 in the United States and six in China.[15] Environmental stewardship became a business trend, just as consumer electronics had been decades earlier.

Also in 2019, Shinjiro Koizumi, Japan's young and charismatic environment minister, pledged to make the government's fight against climate change sexy. "In politics there are so many issues, sometimes boring. On tackling such a big-scale issue like climate change, it's got to be fun, it's got to be cool."[16] After Yoshihide Suga retired in 2024, Koizumi, the thirty-seven-year-old son of former prime minister Junichiro Koizumi, succeeded the elder statesman, bringing a rare youth and vigor to Japanese politics. However, his support of twenty-two new coal plants when he was environment minister haunted his first term in office.[17]

On a visit to his birthplace in Yokosuka, a city of 400,000 on the southern tip of Tokyo Bay, Koizumi, his wife, Christel Takigawa, and his son were confronted by locals who were suing his government for sidestepping full environmental review of the two plants being built in their neighborhoods. Yellow banners saying, "Why coal? Why now?" hung from hundreds of buildings. By the end of the day, the prime minister was asking the same question himself. He returned to work the next day more determined to protect Japanese coastal communities like his hometown from the ravages of a heating climate.

His first action was to rescind the offer of $5 billion of Japanese public financing for coal projects being built in the developing world, including Southeast Asia's largest coal-fired power plant in Indonesia, as well as high-polluting plants in environmentally sensitive areas in Vietnam and the Philippines.[18] He also started working with corporate leaders to develop a BlueGreenTech policy that focused on clean energy innovation and carbon capture and storage infrastructure. Some were homegrown Japanese technologies, since Japan had once boasted the most worldwide patents in the renewable energy sector but had been slow to commercialize them.[19] He directed massive government investment toward the sector, and Japan's lackluster climate commitments were suddenly supercharged.

Japan's energy landscape changed through the 2020s and 2030s. The six countries that belonged to the One Planet Sovereign Wealth Fund Working Group decided on an "each one, reach one" strategy to grow their club of wealthy nations.[20] Norway reached out to the Japanese, whose sovereign wealth fund had surpassed the Norwegians' to become the world's largest, with $1.5 trillion invested. The Saudi Public Investment Fund reached out to China, which began investing heavily in infrastructure projects around the world. Together, these investments provided the boost emerging energy companies needed to win the market, as solar had done in the 2010s. With decreasing demand for fossil fuels and genuine, low-cost alternatives, the remaining major drilling countries—except Russia—also restructured their wealth funds away from gas, oil, and coal for good.

At the same time, global interfaith leaders continued to act. First, they agreed on the Religious Forestry Standards, which a number of faiths had worked to develop with the Worldwide Fund for Nature and the Forest Stewardship Council in the 2010s, since the world's religions own vast tracts of land all over the world.[21] It brought 5 percent of the world's commercial forests up to the highest sustainability standards. Then the faith leaders moved from discussing beliefs to talking business: how to bring their land holdings and capital into alignment with their shared ecological values. As Martin Palmer of the Alliance of Religions and Conservation said in 2017, "We have known for some time what the faiths were against in their investments. But now we—and they—have a much better idea of what they are for."[22]

This small initial meeting led to the 2024 launch of the WorldCare Standard, a voluntary but later widely adopted set of guidelines that rejected dirty energy investments and moved nearly 10 percent of the global investment portfolio into B corporations, nonprofit organizations, and other social impact initiatives. Jews, Christians, and Muslims called these investments "creation-positive," while non-theist religions referred to them as "nature-aligned," but it all meant the same thing in the real world. One trillion dollars of church-, temple-, synagogue-, and mosque-owned assets shifted over the following decade, pushing those profit-making enterprises that were not acting responsibly to clean up their game. The world's religions—those barges of culture that turn slowly but influentially—had taken centuries to more fully inhabit the principles of their founders. But once the existential threat to the earth became imminent, they finally took up the mantle of moral leadership.

Back in Ise, the Shinto priests performed their venerations and made their offerings to Amaterasu as they always had. Visits to shrines steadily increased as people found more meaningful ways to spend their time and money.[23] The Jinja Honcho started secret discussions about what would be done with the thousands of shrines, especially Geku—and possibly even Naiku—if seas continued to rise. They scouted sites in the sacred forests further inland, places where shrines could be rebuilt if necessary in forty years, or two hundred, or a thousand. As seas began to rise more quickly in mid-century, their planning became more urgent and public. Now, with sea levels nearly two feet higher than they were at the century's start, Shinto is reinventing its relationship with the *kami* in response to changes far beyond any one culture's control.

WHAT'S NEXT?

Marina Psaros

Utopia is the process of making a better world. —Kim Stanley Robinson

When I began working on climate change planning in 2007, photos of forlorn-looking polar bears crouched on thin ice seemed to accompany nearly every news story about the looming crisis. Those sad polar bears may have made for arresting visuals, but they didn't convey that climate change would also hit much closer to home. At the time, almost no one was talking about or planning for California's climate change risks, like the inequitable health impacts of flooding in San Diego's low-income neighborhoods, or the future of the Bay Area's coastal real estate, or the loss of SoCal's beaches. My work through the National Oceanic and Atmospheric Administration (NOAA) was to bridge this gap by helping local communities start planning for climate change impacts.

My colleagues and I believed the most efficient way to do this would be to get directors at local government agencies to make it a top priority for their departments. So we translated what the national science agencies were discovering into community policy and planning tools. After a year or so of working with these department heads, we started getting feedback like this: "Thank you, this is a critical issue, but I'm not the changemaker here, You need to train up my staff. They're the ones who really matter because they're the ones who are actually writing the plans and the ordinances that determine how our communities will look and function."

We repeated the process, this time bringing climate science to staff-level planners and policy analysts. At the end of the process, we were again met with "Thank you, this is a critical issue, but I'm not the changemaker here. You need to convince the elected officials, because they're the ones who approve the plans and the ordinances that I write." Rinse and repeat, more technical assistance programs and workshops, and once again we heard, "Thank you, this is a critical issue, but I'm not the changemaker here. You need to convince the public, because I respond to what my constituents tell me is a priority."

And "the public"? They, of course, are looking to the scientists, planners, and policymakers for the answers. After witnessing this passing of the buck for over a decade, I've come to the conclusion that we need to stop looking around for someone else to solve the problem and start taking action within our own spheres of influence. For "the public"—you and me—this means making different decisions about how we choose to spend our money, our time, and our energy. As consumers, community members, and citizens, we each inhabit multiple networks in which we can act as change-makers. The following ideas are not the comprehensive plan to stop climate change; rather, they are actions you can take as an individual to become part of the collective action that is necessary to solve the crisis.

What You Can Do as a Consumer

In the spring of 2020 as the COVID-19 pandemic upended the lives of billions of people, we saw that the collective impact of individual actions—what we buy, eat, and do—swiftly and directly influenced the environment. From clean skies over Los Angeles and Shanghai to wildlife walking city streets and a temporary 17 percent reduction in global carbon dioxide emissions, stay-at-home orders demonstrated clearly that our planet is resilient. We also learned in that painful period that humanity is capable of rapid changes that alter our day-to-day lives. Our challenge now is to find ways to sustain massive change while maintaining quality of life. One place to start is how we spend our money.

American household consumption was responsible for 20 percent of worldwide greenhouse gas (GHG) emissions at the turn of this century, so clearly we can exercise a lot of power through what we choose to buy. Plastics are a good place to start to change your consumption: a few simple actions include carrying a reusable water bottle instead of drinking from plastic, replacing your plastic Tupperware with glass or metal containers, and using beeswax-coated cloth instead of plastic wrap. A quick internet

search will offer alternatives for nearly every plastic item that you use.

Next, you can make sure that your food choices are climate smart. Industrially produced animal meat contributes to deforestation, overuse of fertilizers and antibiotics, and potent gas emissions. If you do choose to eat meat, you could hunt or raise your own, or buy from local operations that harvest meat sustainably. FoodPrint (www.food print.org) and Seafood Watch (www.sea foodwatch.org) publish frequently updated information about more sustainable meat choices for specific geographic locations.

You can also divest yourself from fossil fuels. Become car-free or car-light and support public transportation by using it and voting to pay for it. Transportation accounts for almost 30 percent of GHG emissions in the United States; over 90 percent of the fuel used is petroleum-based. Continuing to depend on a car for daily life perpetuates inequitable, unsafe, and environmentally disastrous land use patterns. Before you invest in a new car—even an all-electric one—consider if moving closer to your work or public transit is feasible and perhaps a better investment of your money and time. In your home, you can opt in to 100 percent GHG-free electricity from your local utility and use rebates and incentive programs to install solar panels and replace your gas-powered appliances (like clothes dryers, heaters, and stoves) with electric ones.

What You Can Do as a Citizen

You can keep climate change on the policy radar by letting your elected officials know that limiting its impacts is a priority for you, their constituent. Join the local chapter of your favorite environmental advocacy organization so that they can tell you when and how to call your local, district, state, and congressional representatives. Calling, writing emails, participating in surveys, attending meetings, and otherwise directly engaging with your elected officials are actions that carry more weight than liking something on social media or signing form petitions. When election season rolls around, ask your local candidates to speak directly about their climate change policies in debates, on their websites, and in the media. The climate emergency should be a non-partisan issue, and recent polling finds that younger voters in both parties share concern about climate change.

You can serve on a citizen oversight committee, board, or office that makes decisions about energy and land use for your locality. Many cities and states are now moving faster and further than the federal government in enacting climate-friendly policies, from 100 percent renewable energy in Greensburg, Kansas, to state-funded buyout programs for flood-prone property in Louisiana, New Jersey, and South Carolina to protecting migration corridors for wildlife in New Mexico. All of these policies were enacted through the public process by people who showed up and made their priorities a reality.

What You Can Do in Your Communities

Each of us belongs to many different communities: we belong to families, schools, workplaces, houses of worship, gyms, clubs, neighborhoods, online spaces. In every one of these communities, you can engage in respectful dialogue about the kind of future that you want, and that community can be a part of building it. As a planner and a public servant, I have participated in hundreds of these dialogues over the years and I have seen that starting with shared experiences and goals can help people stay productive and collaborative despite their differences. For example, you might not share religious or political affiliation with your neighbors, but you do share the same air quality, water and food sources, municipal services, and weather—and it's pretty likely you would all agree that you want those to function well.

No one wants their town to be consumed by rising seas or for the oceans to be vast expanses of dead zones, so start there. Conversations with people in your various communities can start from a place of inquiry and a spirit of collaboration. You can talk about climate change as it relates to the shared values and norms of the community in which you both belong. Can your PTA help implement a climate change curriculum at your kids' school? Has your workplace conducted an energy efficiency audit? Has your gaming group ever participated in a Green Game Jam? Does your town have climate-friendly development policies? The problems that we are trying to solve are complex, and the most powerful thing we can do is take the next step in front of us.

Let's begin.

TOWARD TRANSILIENCE

Christina Conklin

As I was finishing this manuscript in early 2020, a contagion began spreading around the world like wildfire, upending economies and lives. COVID-19 has since rewritten reality overnight, at least for a while, showing how connected and vulnerable we all are. The pandemic has revealed compassion, solidarity, and hopeful action in ways that were inspiring and instructive to those of us working to contain the climate crisis; most people changed their behaviors to protect themselves and others, and some even welcomed the wakeup call to greater self-awareness that accompanied the hardships we never asked for or wanted. COVID-19 also unleashed waves of fear, anger, denial, and depression, and many of us are still going through these various stages of grief.

Grief is our natural response to traumatic loss. Experts who work with people feeling either personal or ecological grief urge us not to turn away from sorrow, but to use our grief to transform our most painful experiences into more meaningful relationships with each other and the rest of nature. These are skills best practiced in community, which brings people together in a political act of sol-idarity, rather than leaving them in isolation. In bearing witness together to the very real suffering caused by both the pandemic and the climate emergency, we can strengthen our ability to stay present with a breaking world and then take bold action on its behalf. This is challenging, deepening work. It accepts impermanence and uncertainty as its foundation; it understands that we cannot know the outcome, only our role in the process; it pays attention to whatever arises, in order to live with more compassion, flexibility, and equanimity. Grief is a gateway to growth. As the poet Rainer Maria Rilke wrote, "Let this darkness be a belltower."[1]

COVID-19 was a clarion call for me, telling me that the climate crisis, though enormous, is not the only challenge we face as a species. In fact, a number of scholars and activists are working on the intersectional nature of many different global stressors that are now interacting in ways never seen before. For instance, the pandemic pushes on economic inequality, hunger, the ozone layer, water quality, and election outcomes. Some thinkers call this the "global polycrisis" or the "human predicament." Others call it the study of existential risk, likening it to the state of constant insecurity that is usually only common in times of war and already common in many places in the Global South. What is new in the twenty-first century is the complexity of the web, the speed at which change cascades through the systems, and the fact that disturbances can originate at so many points. We have entered a more contingent time in human history. Some call it brittle, but we can also see it as fluid.

Understanding ourselves as part of and entirely embedded within the living world is the first step toward creating the sustained healing we need to address many of our global challenges. This would be a massive paradigm shift away from long-held ideas of human superiority and dominion over nature. Fortunately, this shift has been in the works for decades. James Lovelock's Gaia Theory in the 1960s describes the world as a superorganism that adapts over time, and many recent philosophers—Gilles Deleuze, Bruno Latour, Michel Serres, Isabelle Stengers, Pieter Sloterdijk, Donna Haraway, and Timothy Morton, to name a few—have reimagined our place in a more-than-human, dynamic, process-driven world-system.[2] For people concerned with the health and viability of life, this has been the most important intellectual shift of the past century. It puts us at the crossroads of a new worldview that both better reflects scientific reality and, according to Thomas Kuhn, author of *The Structure of Scientific Revolutions*, will likely lead to an unprecedented flow of ideas and creativity.[3]

Even when leadership lags, all our actions matter. Some researchers who have studied societal shifts have concluded that only a small minority of people need to commit to making change for a paradigm shift to take root.[4] This is very good news. According to a recent poll, more than a third of Americans think climate change is indeed a "crisis" that will require sacrifices. As Katherine Hayhoe, the influential climate scientist and advocate at Texas Tech University, likes to say, "We already have the values we need."[5] We just need to start seeing with new eyes.[6]

In addition to learning to be resilient—recovering from traumatic events, or making incremental changes—we must find ways to become *transilient*—that is, to make the radical shifts required for future generations to survive and thrive. Transilient action takes imaginative leaps. A transilient future includes institutions and governments that divest from fossil fuels in order to fund green-energy infrastructure. It includes private and public investment in developing countries so that they can leapfrog over coal, gas, and oil and skip directly to solar, wind, and wave power.[7] It includes communities, led by their youngest members, staging protests and boycotts to upend the standard talking points of politicians and the media. It includes small groups experimenting with alternative governance and economic systems to grow a million ecological civilizations of biological and cultural diversity. It requires that countries come together to sign an international treaty on carbon pricing that will create world-saving accountability among nations.

To become transilient, we must acknowledge the damage we have already done and the irreversible changes that will roll out for decades, regardless of our current actions. We must shape equitable, flexible new systems and structures that can better adapt to the changes that are coming. And we must invent and invest in the means to pull carbon out of the atmosphere and ocean in order to move our climate back toward dynamic equilibrium. A millennium from now is unimaginable in so many ways, but it is the blink of an eye in terms of climate, evolution, and natural systems. The generations alive today will play a large part in determining what that planet will look like.

So many of the research papers I have read conclude with climate scientists saying, in language tinged with despair, that the only possible solution to the losses they are observing is for societies to *reverse global warming*.[8] Experts say the 2020s are our last best chance to put the genie back in the bottle.

We must start before we feel ready and keep going even when we are tired.

We come from the sea. We are fish out of water. Salt water runs through our veins, and most of the oxygen we breathe is pro-
duced by marine microorganisms. We are in an intimate, elemental relationship with the ocean that sustains us and binds each to all. Returning the ocean's body—and our own— to health is ultimately about writing a new story about our place in the world, and then becoming characters in it.

What will our descendants say of us in a hundred years, in a thousand? In the words of Jonas Salk, are we being good ancestors? In the geologic future, our era will show up as a layer of sediment made from steel, concrete, and plastic. What layer will come after it? In writing this book, I have learned so much about people who are already helping to mend the fabric of the planet, even as others tear it further. There will be suffering in the decades to come, especially for the young, old, poor, marginalized, and vulnerable— and there will also be tremendous opportunity for creative engagement in cultural transformation. What we do, and when, matters.

ACKNOWLEDGMENTS

As with all books, *The Atlas* has been a team effort. It would not exist without the steady, patient hand of our editor Ben Woodward and the creativity and dedication of the editorial and production staff at The New Press. We would also like to thank our agent, Carole Jelen, for helping us navigate the publishing process. The resourceful and good-humored research assistance of Sam Leander and Raven Chalif were indispensable.

Marina—I am grateful to belong to the community of climate change researchers, policymakers, planners, and educators whose professional lives are dedicated to expanding sustainability and stewardship. Many from this community have contributed to *The Atlas* by sharing their work, reviewing ours, and thinking through the hard stuff with me: Elizabeth Bagley, Beatriz de la Torre, Captain Dennis, Ed Carpenter, Ellie Cohen, Dave Eslinger, Carrie Grassi, Robin Grossinger, Brigid McCormick, James Morioka, Ronald Osinga, Sara Polgar, Lucy Reading-Ikkanda, Dan Rizzo, William Skirving, Drew Talley, Joelle Tirindelli, and Christine Whitcraft, to name just a few. Thanks also to my former MIT advisor Larry Susskind for inspiring me in this field and for providing the book's context in the foreword.

The seed of *The Atlas* grew from this fertile soil: the King Tides Project, Youth Exploring Sea Level Rise Science (YESS), and Our Coast, Our Future; Adina Abeles, Sara Aminzadeh, Lauren Armstrong, Patrick Barnard, Dani Boudreau, Linda Chilton, Juliette Finzi Hart, Annie Frankel, Kristen Goodrich, Deborah Hirst, Kurt Holland, Sarah Hutto, Kelley Johnson, Jack Liebster, Rebecca Lunde, Gwen Noda, Heidi Nutters, Hilary Papendick, John Rozum, Susan Schwartzenberg, Heather-Lynn Remacle, Marta Smith, Bobbak Talebi, and Alex Westhoff. Thank you to Jed Bickman, who originally approached me about writing this book and who has continued to provide encouragement and perspective even after leaving The New Press.

Finally, my love and appreciation to the family and friends who've been with me on this epic quest: Arah Schuur—the most steadfast of writing and prodding companions, Rosaclaire Baisinger, Laura Bell, Patricia Chan and John Lyng, Michelle Corral, Jake Donham, Alexis Minsloff, Michelle Park, Mary Powell, Chris, Derrick, Heather, and Marylin Psaros, Jackie Randazzo, Jen Reck, Marc Ritter, Traci Ruble, my CORO Women in Leadership cohort, my Michigan family, and last but never least, Lukas, Ada, and Kai.

Christina—I send my deep love and gratitude to my family, who get me and make me laugh. Richard, my co-navigator and anchor, it has always been you. Gemma, you are my bright, shining star with a heart as big as the sun. Will, you're the smartest, kindest guy I know, and your imagination inspires me every day. Deep thanks also to my parents, sister, and extended family, who have always been there for me with love and support. My dear friends have been the best cheerleading squad I could have hoped for these past four years: my Rising Strong crew, my Half Moon Bay writing group, Dayna, Diana, Dover, Greet, Lori, and Sasha. You're the best.

A number of writers sparked my interest in learning and writing about the crisis in the ocean, especially Elisabeth Mann Borgese, author of *The Drama of the Oceans*, and Rachel Carson, who helped me hew to both science and poetry in my work. I am grateful for their legacy.

I would like to thank many colleagues for taking time to share their experiences and knowledge with me, and for reviewing my chapters. Martine van den Heuvel-Greve and her colleagues at Wageningen University; Jacqueline Evans, Gustaaf Halegraeff, Michelle Bender, Will Travis, Lynn Englum, Daniel Zarrilli, Ayasha Guerin, Klaus Jacob, Peter Helmlinger, Liz Koslov, Stan Sze, Fiona McCluney, Alyson Santoro, John Lambert, Shahnoor Hasan, Laura Manston, Jennifer Haigh, and Peter Behrens. Lane Robson, Cindy Handler, and Sarah Schilbach helped me flesh out the "ocean as body" metaphor. Debbie Ebanks and Lisa Kairos offered invaluable feedback on the artwork, and Erin Higgins helped greatly in preparing the image files. Your collective expertise has enriched the texture and content of this book.

Special thanks to my mentors at California College of the Arts, who introduced me to so many great philosophers and thinkers. And to the writing and art residencies that provided me with amazing opportunities to create, travel, and collaborate, namely the Arctic Circle residency in Svalbard, the Mesa Refuge in Point Reyes Station, California, and the Banff Center writing retreat. Lastly, I am grateful to my many collaborators in the art and activism fields who are thinking with me about the new paradigms and systems we can collectively create.

NOTES

INTRODUCTION

1. For an introduction to these ideas, see George Sessions, editor, *Deep Ecology for the Twenty-first Century: Readings on the Philosophy and Practice of the New Environmentalism* (Boston: Shambhala Press, 1995); Stephen Jay Gould, *Time's Arrow, Times Cycle: Myth and Metaphor in the Discovery of Geological Time* (Cambridge, MA: Harvard University Press, 1988); deepadaptation .ning.com; and Jem Bendell, *Deep Adaptation: A Map for Navigating Climate Tragedy*, Institute for Leadership and Sustainability, University of Cumbria, UK, www.iflas.info, July 27, 2018.

2. For more on the history and variety of map projections, see www.kartograph.org/ showcase/projections/#ortho.

3. For Christina, this approach was inspired by Raymond Williams's classic text, *Keywords: A Vocabulary of Culture and Society*, revised edition (New York: Oxford University Press, 1983).

4. More on these ideas in Bruno Latour, *We Have Never Been Modern* (Cambridge, MA: Harvard University Press, 1993).

5. Conversation between Christina Conklin and Dr. Lane Robson, December 10, 2019.

PART I: CHANGING CHEMISTRY

1. Catherine Jeandel, "Seawater: A Chemical Solution," *The Ocean Revealed*, 2017, 60, www.euromarinenetwork.eu/system /files/2017/The_ocean_revealed_ENG.pdf.

2. Vladimir Vernadsky, a Russian scientist, founded the field of biogeochemistry in his 1926 book, *The Biosphere*, in which he proposed that the earth is a living system. Vladimir I. Vernadsky, *The Biosphere* (New York: Springer-Verlag), 1998.

3. See also Gideon Henderson, "Ocean, Trace Element Cycles," *Philosophical Transactions of the Royal Society A: Mathematical, Physical and Engineering Sciences* 37(2081), November 28, 2016, www.ncbi.nlm.nih .gov/pmc/articles/PMC5069534; Benjamin Wolfe, "Why Does the Sea Smell like the Sea?," *Popular Science*, August 19, 2014.

4. Rob Monroe, "How Much CO2 Can the Oceans Take Up?," Scripps Institution of Oceanography Measurement Notes, July 3, 2013, scripps.ucsd.edu/programs/keeling-curve/2013/07/03/how-much-co2-can-the-oceans-take-up, and IPCC, "Choices made now are critical to our ocean and biosphere," www.ipcc.ch/2019/09/25/srocc-press-release.

5. Ove Hoegh-Guldberg et al., "The Ocean," *Climate Change* 2014, 1675, www.ipcc .ch/site/assets/uploads/2018/02/WGI IAR5-Chap30_FINAL.pdf.

6. Peter Harris, "Climate Change Lags in the Ocean: What It Means for Coral Reefs and Our Grandchildren," https://news.grida.no /climate-change-time-lags-in-the-ocean-what-it-means-for-coral-reefs-and-our-grand children. Nicolas Gruber, "Warming up, turning sour, losing breath: Ocean biogeochemistry under global change," *Philosophical Transactions of the Royal Society A: Mathematical, Physical and Engineering Sciences* 369 (1943): 1980–96.

7. Maia Szalavitz, *Unbroken Brain: A Revolutionary New Way of Understanding Addiction* (New York: St. Martin's Press, 2016).

8. Correct as of October 2020. Amendment to the Montreal Protocol on Substances that Deplete the Ozone Layer, Kigali, October 15, 2016, United Nations Treaty Series, no. 26369, available from: treaties.un.org/doc/Publication/MTDSG/Volume%20II/Chapter%20XXVII/XXVII-2-f.en.pdf.

KURE ATOLL, HAWAI'I: PLASTIC, PLASTIC EVERYWHERE

1. Timothy Morton, *Being Ecological* (Cambridge, MA: MIT Press, 2018), 76.

2. This chapter was inspired by Pietra Rivoli's book *The Travels of a T-Shirt in the Global Economy: An Economist Examines the Markets, Power and Politics of the World Trade*, 2d ed. (New York: Wiley, 2009).

3. See www.csiro.au/en/News/News-releases/2018/How-much-plastic-does-it-take-to-kill-a-turtle, www.wwf.org.au/news/blogs/plastic-pollution-is-killing-sea-turtles-heres-how#gs.257h7i; and www.seeturtles.org/ocean-plastic.

4. Ocean Unite, "Key Issues: Marine Plastic Pollution," www.oceanunite.org/issues/marine-plastic-pollution.

5. A fascinating, comprehensive paper on where plastics in the ocean originate is Jenna Jambeck, Roland Geyer, et al., "Plastic Waste Inputs from Land into the Ocean," *Science* 347: 6223.

6. As the world's largest textile producer, China creates 70 percent of the world's polyester. Torry Losch, "How the Polyester Yarn Supply Chain Is Impacted by the US Trade War with China," September 21, 2018, www.servicethread.com/blog/how-the-polyester-yarn-supply-chain-is-impacted-by-the-u.s.-trade-war-with-china-part-one. Beverley Henry, Kirsi Laitala, and Ingun Grimstad Klepp, "Microfibers from Apparel and Home Textiles: Prospects for Including Microplastics in Sustainability Assessment," *Science of the Total Environment* 652 (2019): 483–94; Cal Recycle, "Textiles," updated December 27, 2019, www.calrecycle.ca.gov/reducewaste/textiles; www.epa.gov/energy/greenhouse-gas-equivalencies-calculator.

7. Alexandra Martins and Lucia Guilhermino, "Transgenerational effects and recovery of microplastics exposure in model populations of the freshwater cladoceran *Daphnia magna* Straus," *Science of the Total Environment* 631 (2018): 421–8; Shima Ziajahromi et al., "Impact of microplastic beads and fibers on waterflea (*Ceriodaphnia dubia*) survival, growth, and reproduction: Implications of single and mixture exposures," *Environmental Science & Technology* 51(22): 13397–406; "Nanoplastics Accumulate in Marine Organisms and May Cause Harm to Aquatic Food Chains," *Science Daily*, May 1, 2018. For more on their effects on marine life, see Maria C. Fossi and Letizia Marsili, "Effects of endocrine disruptors in aquatic mammals," *Pure and Applied Chemistry* 75(11–12): 2235–47, and Guy Linley-Adams, "Cetaceans and Endocrine Disruptors," www.charlie-gibbs.org/charlie/NEA_Web site/Publication/briefings/Cetaceans.pdf.

8. These are common industrial chemicals and are among the most toxic in the global environment. More on their health effects is available on the U.S. Agency for Toxic Substances and Disease Registry, www.atsdr.cdc.gov/substances/index.asp.

9. More information can be found in "The New Plastics Economy: Rethinking the Future of Plastics," *World Economic Forum*, 2016; and Andrea Thompson, "Solving Microplastic Pollution Means Reducing, Recycling—and Fundamental Rethinking," *Scientific American*, November 12, 2018; Evan Lubosky, "Tracking a Snow Globe of Microplastics," *Oceanus*, December 10, 2018.

10. Stockholm Resilience Center, "The Nine Planetary Boundaries," Stockholm University, www.stockholmresilience.org/research/planetary-boundaries. See also information on the 2001 Stockholm Convention on Persistent Organic Pollutants, www.epa.gov/international-cooperation/persistent-organ ic-pollutants-global-issue-global-response.

11. Read more on the REACH program of the European Commission at https://ec.europa.eu/environment/chemicals/reach/reach_en.htm.

12. Benzene, glyphosate, and dioxins and other petrochemicals all cause blood cancer. Information on cancer clusters and a partial list of cancer-causing chemicals can be found at www.atsdr.cdc.gov/emes/public/docs/Chemicals,%20Cancer,%20and%20You%20FS.pdf; www.myeloma.org/blog/dr-duries/toxic-exposures-unleashed; www.cancer.org/cancer/cancer-causes/benzene.html.

13. Emily J. North and Ralph U. Halden, "Plastics and Health: The Road Ahead," *Review of Environmental Health* 28(1): 1–8.

14. For commentary on these green plastics, see Adam Lowry, "Compostable and 'Biodegradable' Plastics Provide False Sense of Responsibility," *TreeHugger*, September 15, 2009, www.treehugger.com/sustainable-product-design/compostable-and-biodegradable-plastics-provide-false-sense-of-responsibility.html; and Renee Cho, "The Truth About Bioplastics," *State of the Planet*, December 13, 2017, blogs.ei.columbia.edu/2017/12/13/the-truth-about-bioplastics.

15. See fact sheet at www.plasticsindustry.org/factsheet/california.

16. More information on "crade-to-cradle" production in Michael Braungart and William McDonough, *Cradle to Cradle: Remaking the Way We Make Things* (New York: North Point Press, 2002); E. Moss, A. Eidson, and J. Jambeck, "Sea of Opportunity: Supply Chain Investment Opportunities to Address Marine Plastic Pollution, Encourage Capital," on behalf of Vulcan, Inc., New York, February 2017, www.encouragecapital.com/wp-content/uploads/2017/03/Sea-of-Opportunity-Plastics-Report-Executive-Summary.pdf; and "The New Plastics Economy: Rethinking the Future of Plastics," Ellen MacArthur Foundation, 2016, www.ellenmacarthurfoundation.org/assets/downloads/EllenMacArthur Foundation_TheNewPlasticsEconomy_Pages.pdf.

17. Most "recycled" plastics contain mostly virgin base stock and only a small percentage of recycled content. Teijin Frontier Company, "Eco Circle™ Fibers," www2.teijin-frontier.com/english/sozai/specifics/ecopet-plus.html; Umair Irfan, "The Race to Save the Planet from Plastic," *Vox*, May 15, 2019, www.vox.com/the-highlight/2019/4/9/18274131/plastic-waste-pollution-bacteria-digestion.

18. The number now is 90 pounds but is sure to rise. Eric Beckman, "The World of Plastics, in Numbers," *The Conversation*, August 9, 2018, www.theconversation.com/the-world-of-plastics-in-numbers-100291.

19. Roland Geyer, Jenna Jambeck, and Kara Lavender Law, "Production, Use, and Fate of All Plastics Ever Made," *Science Advances* 3(7).

20. China is responsible for 28 percent of the world's mismanaged plastic waste, so implementing best waste management practices, as in the United States and Europe, would make a substantial difference. Hannah Ritchie and Max Roser, "Plastic Pollution," https://ourworldindata.org/plastic-pollution. See also Geyer et al., "Production, Use, and Fate of All Plastics Ever Made."

21. Already 100 million metric tons of plastic waste is generated annually by people living in the coastal zone. Increased production, even when combined with better waste management, could actually increase net pollution. Ocean Conservancy, "Fighting for

Trash Free Seas," www.oceanconservancy.org/trash-free-seas/plastics-in-the-ocean.

22. These are among the many new fibers that already exist and are currently scaling up. "Biosynthetics: When Synthetic Doesn't Mean Plastic," *Common Objective*, April 10, 2019, "Bigger proteins, stronger threads: Synthetic spider silk: Engineering scientists use bacteria to create biosynthetic silk stronger and more tensile than before," *Science-Daily*, August 21, 2018.

THE ARABIAN SEA: REGIME SHIFT

1. Kennedy Warne, "The Seas of Arabia," *National Geographic*, March 2012.

2. See "Noctiluca Blooms in the Arabian Sea," https://helgagomes.com/2014/01/31/arabian-sea/ and https://williamkusterldeo.wixsite.com/ldeogoesgomeslab.

3. "Shift in Arabian Sea Plankton May Threaten Fisheries," September 9, 2014, https://www.earth.columbia.edu/articles/view/3189.

4. Natasha Vizcarra, "Winter Blooms in the Arabian Sea," https://earthdata.nasa.gov/learn/sensing-our-planet/winter-blooms-in-the-arabian-sea, October 13, 2015.

5. An excellent overview of the topic is Denise Breitburg, "Declining Oxygen in the Ocean and Coastal Waters," *Science* 359 (2015): 6371.

6. Queste had to send in remote vehicles because it has been too dangerous to do research in these waters for decades, hindering our ability to protect what we cannot even study. See https://phys.org/news/2018-04-dead-zone-underwater-robots-gulf.html and Bastien Queste et al., "Physical Controls on Oxygen Distribution and Denitrification Potential in the North West Arabian Sea," *Geophysical Research Letters* 45(9): 4143–52, May 16, 2018.

7. Dipani Sutaria et al., "Baleen Whale Records from India," International Whaling Commission, 2017, https://arabianseawhalenetworkdotorg.files.wordpress.com/2017/05/sc_67a_cmp_03_rev1_baleen-whale-records-from-india.pdf.

8. For numbers caught by each species and nation, see interactive map at www.seaaroundus.org/data/#/lme. Rima Jabado et al., "Troubled waters: Threats and extinction risk of the sharks, rays and chimaeras of the Arabian Sea and adjacent waters," *Fish and Fisheries* 19(6): 1043–62, August 15, 2018, www.sharkconservationfund.org/drivers-of-the-crisis/; Joshua Learn, "Arabian Sea Sharks May Be the Most Threatened in the World," https://oceana.org/blog/arabian-sea-sharks-may-be-most-threatened-world, December 12, 2018; and https://arabianseawhalenet work.org.

9. Mike Gaworecki, "Ocean warming projected to accelerate more than four-fold over next 60 years: Study," *Mongabay*, January 10, 2019.

10. Tim Craig, "On the shores of the Arabian Sea, pollution erodes a way of life," *Washington Post*, March 15, 2015; Prabu Pingali, "Green Revolution: Impacts, limits, and the path ahead," *PNAS* 109 (31), July 31, 2012.

11. See www.stockholmresilience.org/research/planetary-boundaries/planetary-boundaries/about-the-research/the-nine-planetary-boundaries.html.

12. See Denise Breitburg and Reinette Biggs, Stephen Carpenter, and William Brock, "Turning Back from the Brink: Detecting an Impending Regime Shift in Time to Avert It," *PNAS* 106 (3): 826–31.

13. P.F. Ricci and H. Sheng, "Benefits and Limitations of the Precautionary Principle," *Encyclopedia of Environmental Health*, 2011, 276.

14. David Kriebel and Joel A. Tickner, et. al. "The Precautionary Principle in Environmental Science," *Environmental Health Perspectives*, 109(9): 871. www.ncbi.nlm.nih.gov/pmc/articles/PMC1240435/pdf/ehp0109-000871.pdf.

15. Khairallah Khairallah, "The Many Challenges of Sultan Haitham," *The Arab Weekly*, January 3, 2020. Two somnolent organizations already exist to supposedly address this issue, Regional Organization for Protection of the Marine Environment (ROPME), established in 1979 among the gulf states, and South Asian Sea Program (SASP), established 1995), but they need a reboot. See http://ropme.org/1_WhoWeAre_EN.clx and www.sacep.org/programmes/south -asian-seas/action-plan.

16. See the "Coming Attractions" section of Paul Hawken, ed., *Drawdown: The Most Comprehensive Plan Ever Proposed to Reverse Climate Change* (New York: Penguin, 2017) 179.

CAMDEN, MAINE: SALT, FAT, ACID, (NO) MEAT

1. Michael Pollan, *Cooked: A Natural History of Transformation* (New York: The Penguin Press, 2013), 20.

2. "Study shows seas are more vulnerable to acidification near coasts and rivers, putting common fish species at risk," *Goldschmidt*, August 17, 2018, https://whiteiron.org /uploads/conferences/28/press_releases /fileNUOeKt.pdf; and Gulf of Maine Research Institute, "Ocean Acidification: A Growing Concern in the Gulf of Maine," May 28, 2014, www.gmri.org/news/way points/ocean-acidification-growing -concern-gulf-maine.

3. An introduction to the study conducted by Cardiff University researchers and published in 2018 in the journal *Earth and Planetary Science Letters* is available on Phys. org, "Ocean acidification to hit levels not seen in 14 million years," https://phys.org/ news/2018-07-ocean-acidification-mil-lion-years.html.

4. The Pacific Marine Environmental Laboratory's Carbon Program has an overview of ocean pH here: www.pmel.noaa.gov/co2 /story/A+primer+on+pH.

5. Intergovernmental Panel on Climate Change, 2019: IPCC Special Report on the Ocean and Cryosphere in a Changing Climate [H.-O. Pörtner, D.C. Roberts, V. Masson-Delmotte, P. Zhai, M. Tignor, E. Poloczanska, K. Mintenbeck, A. Alegría, M. Nicolai, A. Okem, J. Petzold, B. Rama, N.M. Weyer (eds.)]. In press.

6. L. Kwiatkowski and J. Orr, "Diverging seasonal extremes for ocean acidification during the twenty-first century," *Nature Climate Change* 8: 141–45.

7. See Craig Welch, "Sea Change: Vital part of food web dissolving," *Seattle Times*, April 30, 2014, apps.seattletimes.com/reports /sea-change/2014/apr/30/pteropod-shells -dissolving.

8. "C.R. Williams et al., "Elevated CO_2 impairs olfactory-mediated neural and behavioral responses and gene expression in ocean-phase coho salmon (*Oncorhynchus kisutch*)," *Global Change Biology*, March 2019.

9. Amalia M. Harrington and Heather J. Hamlin, "Ocean acidification alters thermal cardiac performance, hemocyte abundance, and hemolymph chemistry in subadult American lobsters Homarus americanus H. Milne Edwards, 1837 (Decapoda: Malcostraca: Nephropidae)," *Journal of Crustacean Biology* 39(4).

10. For up-to-date and historical information on the lobster fishery, see Atlantic Marine Fisheries Commission, "Species: American Lobster," www.asmfc.org/species/american -lobster.

11. Conversation with Marina Psaros, August 2019.

12. World Health Organization, "Global and Regional Food Consumption Patterns and Trends," www.who.int/nutrition/topics/3 _foodconsumption/en/index5.html.

13. This statistic comes from FAO's Food Wastage Footprint Summary Report and FAO's Sustainable Development Goals program.

For more on the issue, *Vox* produced an accessible yet thorough video report in 2017 called "Food Waste Is the World's Dumbest Problem," available online, www.youtube.com/watch?v=6RlxySFrkIM.

14. The U.S. Environmental Protection Agency provides an extensive guide to adaptation strategies at www.epa.gov/arc-x/strategies-climate-change-adaptation.

THE COOK ISLANDS: FEEDING THE FEVER

1. Conversation with Christina Conklin, October 10, 2019.

2. Dr. Wolfgang Losacker, "Ciguatera fish poisoning in the Cook Islands," *Pacific Islands Marine Resources Information System*, Bulletin #2, July 1992, 13–14.

3. G.M. Hallegraeff et al., "Ciguatera fish poisoning," *Manual on Harmful Marine Microalgae*, 2003, 730; Maria Faust, "Ocean Literacy at the Microscopic Level," Smithsonian Museum of Natural History, *The Plant Press* 11(2), April 2008, https://nmnh.typepad.com/the_plant_press.

4. G.M. Hallegraeff, "Harmful Algal Blooms: A Global Overview," *Manual on Harmful Marine Microalgae* (Paris: UNESCO) 33, 2003, 1–22; Mireille Chinain et al., "Update on ciguatera research in French Polynesia," *PC Fisheries Newsletter* 150 (May–August 2016, update October 2016), 42, www.researchgate.net/publication/309418826_Update_on_ciguatera_research_in_French_Polynesia; Mark P. Skinner et al., "Ciguatera Fish Poisoning in the Pacific Islands (1998–2008)," *PLoS Neglected Tropical Diseases* 5(12); Anneke Brown, "One in 100: Coral Reefs in Rarotonga and the World," *Cook Island News*, November 23, 2019, www.cookislandsnews.com/features/weekend/item/75027-one-in-100-the-coral-reefs-of-rarotonga-and-the-world.

5. Stephen Leahy, "Tiny Pacific island nations create world's largest marine parks," *Guardian*, August 30, 2012.

6. Zack Budryk, "UN proposal calls to protect 30 percent of Earth by 2030 as species face extinction," *The Hill*, January 13, 2020.

7. Michael G. Petterson and Akuila Tawake, "The Cook Islands (South Pacific) experience in governance of seabed manganese module mining," *Ocean & Coastal Management* 167(1): 271–87; Royal Society of Chemistry, www.rsc.org/periodic-table/element/25/manganese; Rupert Neate, "Seabed mining could earn Cook Islands tens of billions of dollars," *The Guardian*, August 5, 2013; Andrew Chin, Katelyn Hari, and Hugh Goven, "Predicting the Impacts of Mining Deep Sea Polymetallic Nodules in the Pacific Ocean: A Review of Scientific Literature," Deep Sea Mining Campaign and Mining Watch Canada, May 2020, www.deepseaminingoutofourdepth.org/wp-content/uploads/Nodule-Mining-in-the-Pacific-Ocean-2.pdf, 45.

8. Jacqui Evans in conversation with Christina Conklin, October 3, 2019.

9. Ben Doherty, "Cook Islands: Manager of world's biggest marine park says she lost job for backing sea mining moratorium," *The Guardian*, October 19, 2019.

10. Global Alliance for the Rights of Nature, "Thomas Berry's Ten Principles of Jurisprudence," www.rightsofnature.org.

11. When the law passed, Gerrard Albert, lead negotiator for the Maori, explained: "Rather than us being masters of the natural world, we are part of it. We want to live like that as our starting point. And that is not an anti-development, or anti-economic use of the river but to begin with the view that it is a living being, and then consider its future from that central belief." Dan Cheater, "I am the River, and the River is me: Legal personhood and emerging rights of nature," *West Coast Environmental Law Alert Blog*, March 22, 2018, www.wcel.org/blog/i-am-river-and-river-me-legal-personhood-and-emerging-rights-nature.

Mihnea Tansescu, "When a river is a person: From Ecuador to New Zealand, nature gets its day in court," *The Conversation*, June 19, 2017.

12. Monica Evans, "Cook Islands to grant sea bed mining exploration licenses within a year," *Mongabay*, June 17, 2020, https://news.mongabay.com/2020/06/cook-islands-to-grant-seabed-mining-exploration-licenses-within-a-year.

13. "Predicting the Impacts of Mining Deep Sea Polymetallic Nodules in the Pacific Ocean: A Review of Scientific Literature" provides an excellent overview of the many issues caused by sea bed mining, https://miningwatch.ca/sites/default/files/nodule_mining_in_the_pacific_ocean.pdf.

14. "The Treaty in Brief," New Zealand History, https://nzhistory.govt.nz/politics/treaty/the-treaty-in-brief.

15. Such a union was first proposed in 2003. "Australia floats 'Pacific Union' idea," ABC News, August 17, 2003.

SAN FRANCISCO BAY: SEVEN-LAYER DIP

1. Rebecca Solnit, *Infinite City: A San Francisco Atlas* (Berkeley: University of California Press, 2010).

2. In 1961, three women, Catherine Kerr, Sylvia McLaughlin, and Esther Gulick, who were concerned about the health of the San Francisco Bay, formed an advocacy group that secured major environmental victories for the region and still exists today, www.savesfbay.org.

3. Joseph Geha, "Pact reached to make Newby Island stop dumping pollutants into bay," *San Jose Mercury News*, April 18, 2018. "Milpitas: San Jose Planners Allow Newby Island Landfill Expansion," *San Jose Mercury News*, December 9, 2016, www.republicservices.com/municipality/newby-island.

4. Isabella Isaacs-Thomas, "The House just voted to regulate PFAS. Here's what you need to know," PBS NewsHour, January 10, 2020.

5. Rachel Ross, "What Are PFAS?," *LiveScience*, April 30, 2019, www.livescience.com/65364-pfas.html. Nadia Kounang, "What are PFAS chemicals, and what are they doing to our health?" *CNN* online, February 14, 2019. Per- and polyfluoroalkyl Substances (PFAS) State Legislation, www.ncsl.org/research/environment-and-natural-resources/per-and-polyfluoroalkyl-substances-pfas-state-laws.aspx, March 11, 2020. David Andrews, "FDA Study: PFAS Chemicals More Toxic than Previously Thought," Environmental Working Group, March 9, 2020. Annie Snider, "White House, EPA headed off chemical pollution study," *Politico*, May 14, 2018. For the history of PFAS production and litigation, see "For Decades, Polluters Knew PFAS Chemicals Were Dangerous but Hid Risks from Public," Environmental Working Group, www.ewg.org/pfastimeline/, and Nathaniel Rich, "The Lawyer Who Became DuPont's Worst Nightmare," *New York Times Magazine*, January 6, 2016.

6. See the San Francisco Bay Regional Water Control Board's page on their clean water action plan in the bay: "San Francisco Bay PCBs TMDL Project," updated July 3, 2018, www.waterboards.ca.gov/sanfranciscobay/water_issues/programs/TMDLs/sfbaypcbstmdl.html.

7. One example is in the East Bay town of Pleasanton, which in late 2019 discovered that PFAS levels exceeded state safety thresholds. For a summary of Pleasanton's situation, see Brett Simpson's article "Silicon Valley Suburb Scrambles to Tackle 'Forever Chemicals,'" *Earth Island Journal*, February 28, 2020.

8. Patricia Kime, "Dozens More Military Bases Have Suspected 'Forever Chemical' Contamination," Military.com, April 3, 2020. More on health effects of PFAS at "PFAS: An Overview of the Science and Guidance

for Clinicians on Per- and Polyfluoroalkyl Substances," 2019, www.atsdr.cdc.gov /pfas/docs/ATSDR_PFAS_Clinical Guidance_12202019.pdf, and Agency for Toxic Substances and Disease Registry, Toxicological Profile, www.atsdr.cdc.gov/, 42–44.

9. Ray Kurzweil, "This Is Your Future," Special to CNN, December 26, 2013, www.cnn .com/2013/12/10/business/ray-kurzweil -future-of-human-life/index.html.

10. In 2005 and 2009 studies by the Environmental Working Group, more than two hundred industrial chemicals were found in newborns' umbilical cord blood. More than 180 of the chemicals present are known causes of cancer, birth defects, or brain and nervous system damage. A 2015 study at the University of California at San Francisco found that concentrations of polychlorinated biphenyls (PCBs), organochlorine pesticides (OCPs), polybrominated diphenyl ethers (PBDEs), perfluorinated compounds (PFCs), mercury, and lead were present in umbilical cord blood.

11. "DowDuPont, Chemours, Named in GenX Lawsuit," *Chemical and Engineering News*, October 16, 2017, https://cen.acs.org /articles/95/i41/DowDuPont-Chemours -named-GenX-lawsuit.html.

PART II: STRENGTHENING STORMS

1. Rick Hanson, *Resilient: How to Grow an Unshakeable Core of Calm, Strength, and Happiness* (New York: Harmony Press, 2018).

2. Hurricane Dorian, which devastated the Bahamas in September 2019, was a theoretical Category 6 hurricane with a windspeed of 185 mph. Jeff Masters, "Hurricane Dorian Was Worthy of a Category 6 Rating," *Scientific American*, October 3, 2019.

3. NOAA National Centers for Environmental Information, "Climate at a Glance: Global Time Series," June 2018, www.ncdc.noaa .gov/cag/; John Walsh et al., "Chapter 2: Our Changing Climate," *Climate Change Impacts in the United States: The Third National Climate Assessment* (2014): 19–67.

4. The Geophysical Fluid Dynamics Laboratory maintains a website that summarizes emerging research in hurricane frequency and intensity at www.gfdl.noaa.gov.

5. "Ocean Fact Sheet," United Nations Ocean Conference, June 2017, www.un.org/ sustainabledevelopment/wp-content /uploads/2017/05/Ocean-fact-sheet -package.pdf; Mark Pelling and Sophie Blackburn, *Megacities and the Coast: Risk, Resilience, and Transformation* (New York: Routledge, 2013).

6. U.S. Climate Toolkit, "Social Vulnerability Index," United States Global Change Research Program, August 2, 2019, toolkit .climate.gov/tool/social-vulnerability-index; Hazards and Vulnerability Research Institute, "Social Vulnerability Index for the United States 2010–2014," University of South Carolina College of Arts and Sciences, artsandsciences.sc.edu/geog/hvri /sovi%C2%AE-0.

7. Nicoletta Lanese, "Fight or Flight: The Sympathetic Nervous System," *Life Science*, May 9, 2019; Sarah Klein, "Adrenaline, Cortisol, Norepinephrine: The Three Major Stress Hormones, Explained," *Huffington Post*, April 19, 2013.

8. National Institute of Building Sciences, "Natural Hazard Mitigation Saves: 2017 Interim Report," www.nibs.org/page/ mitigationsaves.

HOUSTON: WE HAVE A PROBLEM

1. Ed Emmett in a PBS interview that aired on February 5, 2018, www.pbs.org/newshour /show/how-decades-of-houston -development-add-up-to-rising-flood-risk.

2. Many of the exports and imports of the

channel are petroleum related, making the Houston Ship Channel the "largest petrochemical manufacturing complex in the Americas," according to "Port Houston: The International Port of Texas," www.porthouston.com.

3. Charise Johnson and Yvette Arellano, "Air Toxics and Health in the Houston Community of Manchester Fact Sheet," Union of Concerned Scientists and Texas Environmental Justice Advocacy Services, June 2016, www.ucsusa.org/sites/default/files/attach/2016/06/ucs-manchester-air-toxics-and-health-factsheet-2016.pdf.

4. National Oceanic and Atmospheric Administration, "U.S. Billion Dollar Weather and Climate Disasters 1980–2020."

5. See Peter Holley, "Why One Expert Predicts a Major Hurricane Hitting Houston Would Be "America's Chernobyl," *Texas Monthly*, August 21, 2020, and Eliza Barclay, "Harvey is part of a pattern of extreme weather scientists saw coming. They're still shocked," *Vox*, August 31, 2017.

6. Roy Scranton, "When the Next Hurricane Hits Texas," *New York Times*, October 7, 2016.

7. Nsikan Akpan, "Hurricane Harvey damaged petrochemical refineries, releasing thousands of pounds of airborne pollutants," *PBS NewsHour*, August 29, 2017.

8. Julianne Crawford, "Environmental Racism in Houston's Harrisburg/Manchester Neighborhood," March 15, 2018, http://bay.stanford.edu/blog/2018/3/15/environmental-racism-in-houstons-harrisburg-manchester-neighborhood.

9. WE ACT for Environmental Justice, "Assisting Congress to Better Understand Environmental Justice," 2013, www.sipa.columbia.edu/academics/capstone-projects/assisting-congress-better-understand-environmental-justice.

10. The Center for Biological Diversity's geospatial analysis is available through their Digital Assets Database at www.biologicaldiversity.org/news/press_releases/2017/harvey-superfund-sites-09-08-2017.php.

11. "Addressing the Needs of Immigrants in Response to Natural and Human-Made Disasters in the United States," American Public Health Association, November 2006.

12. Junia Howell and James R. Elliott, "Damages Done: The Longitudinal Impacts of Natural Hazards on Wealth Inequality in the United States," *Social Problems* 66(3): 448–67, and *There Is No Such Thing as a Natural Disaster* (New York: Routledge, 2006), by Gregory Squires and Chester Hartman, for an in-depth case study of Hurricane Katrina.

13. Howell and Elliott, "Damages Done."

14. Robert D. Bullard and Beverly Wright, *The Wrong Complexion for Protection: How the Government Response to Disaster Endangers African-American Communities* (New York: New York University Press, 2012).

15. WE ACT, "Assisting Congress."

HAMBURG, GERMANY: RIVER CITY AT RISK

1. One of the worst recorded North Sea storms is known as the *Grote Mandrenke* ("Great Drowning of Men") in Low German. On January 16, 1392, a massive storm tide swept across the North Sea and killed an estimated 100,000 people, while also creating the Wadden Sea and sinking the North German city of Rungholt.

2. Munich Re, "50th anniversary of storm surge in Hamburg: Subsequent flood controls prevent billions in losses," February 13, 2012, www.munichre.com/en/company/media-relations/media-information-and-corporate-news/media-information/2012/2012-02-13-50th-anniversary-of-storm-surge-in-hamburg-subsequent-flood-controls-prevent-billions-in-losses.html.

3. Carsten Nesshöver et al., "The science, policy and practice of nature-based solutions:

An interdisciplinary perspective," *Science of the Total Environment* 579 (2017): 1215–27.

4. See Nando Sommerfeldt and Holger Zschäpitz, "Strompreis-Kollaps durch 'Herwart' offenbart Wahnsinn der Energiewende," *Welt*, October, 31, 2017, www.welt.de/wirtschaft/energie/article 170189224/Strompreis-Kollaps-durch -Herwart-offenbart-Wahnsinn-der -Energiewende.html.

5. Two surveys that aim to rank the world's most livable cities are Mercer's Quality of Living Survey and *The Economist*'s Global Livability Ranking. Both companies have included Hamburg in the top twenty of their last five surveys.

6. Nate Berg, "Germany's Designer City: Can government build Hamburg's next hot neighborhood?," *Next City*, December 16, 2013, nextcity.org/features/view/ germanys-designer-city.

7. For more on resilience and trauma, see Steven Hobfoll, "Social and Psychological Resources and Adaptation," *Review of General Psychology* 6(4): 307–24, and Shelley E. Taylor et al., "Psychological resources, positive illusions, and health," *American Psychologist* 55(1): 99–109.

NEW YORK, NEW YORK: CAPITAL OF CAPITAL

1. Conversation with Christina Conklin, New York City, April 10, 2018.

2. Sandy's surge height is often reported as 14 feet (3.5 meters), but that was the measurement from the mean low tide, whereas climate scientists use the mean high tide to calculate storm surge. Ben Strauss et al., "New York and the Surging Sea: A Vulnerability Assessment with Projections for Sea Level Rise and Coastal Flood Risk," *Climate Central Research Report*, April 2014; Jeremy Deaton, "The Curse of 'Green Gentrification,'" *Medium*, January 25, 2018; Madeleine Lopeman, "Was Hurricane Sandy the 100-Year Event?," *State of the Planet*, May 20, 2015, blogs.ei.columbia.edu/2015/05/20/ was-hurricane-sandy-the-100-year-event-2/; conversation between Liz Koslov and Christina Conklin, April 10, 2018.

3. More information on the state's Build It Back program is available at https://storm-recovery.ny.gov/housing/buyout-acquisi-tion-programs. Strauss et al., "New York and the Surging Sea."

4. Liz Koslov in conversation with Christina Conklin, April 8, 2018.

5. "New York City's Roadmap to 80 x 50," Mayor's Office of Sustainability, 2016.

6. Conversation with Christina Conklin, April 8, 2018.

7. Details in "Coastal Protection," A Stronger, More Resilient New York, www1.nyc.gov /assets/sirr/downloads/pdf/Ch3_Coastal _FINAL.singles. "Resilient Edgemere Community Plan," NYC Department of Housing Preservation and Development, 2017. Liz Koslov in conversation with Christina Conklin, April 8, 2018. Lynn Englum in conversation with Christina Conklin, February 24, 2020.

8. David W. Chen, "In New York, Drawing Flood Maps Is a Game of Inches,'" *New York Times*, January 7, 2018; Ike Brannon and Ari Blask, "The government's hidden housing subsidy for the rich," *Politico*, August 8, 2017.

9. Sarah Crean, "No Retreat from the Coastline," *Gotham Gazette*, June 12, 2013, www .gothamgazette.com/environment/4255 -no-retreat-from-the-coastline.

10. Anne Barnard, "The $119 Billion Sea Wall That Could Save New York . . . or Not," *New York Times*, January 17, 2020.

11. The military has been clear-eyed regarding disaster and climate planning for decades, but the Army Corps, as the branch that has the most civilian interface, tends to be chronically underfunded. Brigadier General Peter Helmlinger in conversation with Christina Conklin, March 10, 2018.

12. "Mayor De Blasio and FEMA Announce Plan to Revise NYC's Flood Maps," www.fema.gov/news-release/2016/10/17/mayor-de-blasio-and-fema-announce-plan-revise-nycs-flood-maps, October 17, 2016.

13. "Safeguarding Our Shores: Protecting New York's Coastal Cities from Climate Change," https://comptroller.nyc.gov/reports/safeguarding-our-shores-protecting-new-york-citys-coastal-communities-from-climate-change/, May 9, 2019; Dan Zarrilli in conversation with Christina Conklin, April 10, 2018. See also Andrew Rice, "When Will New York City Sink?," *New York Magazine*, September 7, 2016.

14. Herman Daly, "Growthism: Its Ecological, Economic, and Ethical Limits," www.resilience.org/stories/2019-03-28/growthism-its-ecological-economic-and-ethical-limits.

15. Olga Kharif, "The Return of the Barter Economy, Swapping Eggs for Toilet Paper," *Bloomberg*, March 31, 2020; Kevin Rector, "Bowl of oranges for a bunch of basil: Strapped for cash, Angelenos turn to bartering and sharing," *LA Times*, May 17, 2020.

16. The city has created a rudimentary evacuation map at floodzonenyc.com, but it is far from an action plan.

17. Regional Plan Association, "The Fourth Regional Plan: Making the Region Work for All of Us," November 2017, library.rpa.org/pdf/RPA-The-Fourth-Regional-Plan.pdf, 13, 36.

18. See a range of rewilding ideas at the Welikia Project, https://welikia.org.

19. New York's trains were all above ground until the Great Blizzard of 1888, when the idea of an underground train network seemed like a safer long-term investment: www.history.com/this-day-in-history/great-blizzard-of-88-hits-east-coast.

20. Daniel Zarrilli in conversation with Christina Conklin, April 10, 2018.

SAN JUAN, PUERTO RICO: PODER, DESPACITO

1. Judith Herman, *Trauma and Recovery: The Aftermath of Violence—From Domestic Abuse to Political Terror* (New York: Basic Books, 2015).

2. See quote from Vega Alta mayor Oscar Santiago, as reported in Jose de Cordoba, "Puerto Rico Tallies Up Devastation from Hurricane Maria," *Wall Street Journal,* September 24, 2017.

3. See 2017 Hurricane Season FEMA After-Action Report, July 12, 2018, 20, and GAO's Sept 2018 report "2017 Hurricanes and Wildfires, Initial Observations on the Federal Response and Key Recovery Challenges."

4. In a *New York Times* article, the commanding officer from the National Guard said that he didn't know why the goods were not distributed after they left. The head of the elections commission said that he had repeatedly called officials, but no plans for distribution were made. Frances Robles, "Containers of Hurricane Donations Found Rotting in Puerto Rico Parking Lot," *New York Times*, August 10, 2018.

5. Patrice Taddonio, "Inside the Federal Response to Maria: 'Is This Really the Best FEMA Can Do?,'" *PBS* online, May 1, 2018.

6. "DoD Continues Support in Hurricane-Ravaged Areas," U.S. Department of Defense Newsroom, September 29, 2017, dod.defense.gov/News/Article/Article/1329376/dod-continues-support-in-hurricane-ravaged-areas.

7. See 2017 Hurricane Season FEMA's After-Action Report, July 12, 2018, 20, www.fema.gov/media-library-data/1531743865541-d16794d43d3082544435e1471da07880/2017FEMAHurricaneAAR.pdf.

8. The *New York Times*' methodology was to compare additional deaths in 2017 over deaths in 2015 and 2016, and also to account for outward migration.

Frances Robles, Kenan Davis, Sheri Fink, and Sarah Almukhtar, "Official Toll in Puerto Rico: 64. Actual Deaths May Be 1,052," *New York Times*, December 9, 2017. Harvard's methodology was to survey 3,299 randomly chosen households in early 2018 and extrapolate results for the entire island. Nishant Kishore et al., "Mortality in Puerto Rico After Hurricane Maria," *New England Journal of Medicine* 379(2): 162–70. FEMA's number can be found in "Ascertainment of the Estimated Excess Mortality from Hurricane Maria in Puerto Rico," Milken Institute School of Public Health at The George Washington University, September 24, 2018, 8–14, publichealth.gwu.edu/sites /default/files/downloads/projects/PRstudy /Acertainment%20of%20the%20Estimated %20Excess%20Mortality%20from%20 Hurricane%20Maria%20in%20Puerto%20 Rico.pdf.

9. Doug Mack, "The Strange Case of Puerto Rico: How a Series of Racist Supreme Court Decisions Cemented the Island's Second-class Status," *Slate*, October 9, 2017.

10. World Bank Press Release 2018/136/LAC: "World Bank Provides US$65 million for Dominica's Post-Maria Reconstruction," April 13, 2018, www.worldbank.org/en /news/press-release/2018/04/13/world -bank-provides-us65-million-for-dominicas -post-maria-reconstruction.

11. Carolyn Beeler, "How a Small Caribbean Island Nation Is Trying to Become Climate Resilient," *The World*, March 20, 2020.

12. Rafael Bernal, "Puerto Rico governor asks Trump to consider statehood," *The Hill*, September 19, 2018.

13. "Puerto Rico, the oldest colony in the world, gives the world a master class on mobilization," *Boston Globe*, July 26, 2019.

14. "Puerto Rico's New Push for Statehood, Explained," Alexia Fernández Campbell, *Vox*, October 30, 2019, www.vox.com/identities /2019/10/30/20939916/puerto-rico-state hood-bill-congress.

KUTUPALONG CAMP, BANGLADESH: HUMAN TIDES

1. Tahmima Anam, "The Rohingya crisis is not an isolated tragedy—it's the shape of things to come," *The Guardian*, May 19, 2015.

2. This story is condensed from "Never again, and again and again," *The Economist*, December 8, 2018, 59.

3. "Cyclones and Floods in Myanmar," http:// factsanddetails.com/southeast-asia /Myanmar/sub5_5h/entry-3138.html; "Impact of Climate Change and the Case of Myanmar," Myanmar Climate Change Alliance, www.myanmarccalliance.org/en /climate-change-basics/impact-of-climate -change-and-the-case-of-myanmar/#link5; Khandker Masuma Tasnim et al., "Numerical Simulation of Cyclonic Storm Surges over the Bay of Bengal Using a Meteorology-Wave-Surge-Tide Coupled Model," *Coastal Engineering Proceedings* 1(34): 26; "Myanmar Climate Change Strategy and Action Plan (MCCSAP) 2016–2030," January 19, 2017, www.preventionweb.net/files/65014 _mccsapfebversion.pdf.

4. MCCSAP, 15–16.

5. "The Salty Taste of Climate Change," *Myanmar Times*, September 10, 2019.

6. See "How Climate Change Can Fuel Wars," *The Economist*, May 25, 2019, and Marshall Burke and Solomon Hsiang, "Climate and Conflict," National Bureau of Economic Review, October 2014, 17; C. F. Schleussner et al., "Armed-conflict risks enhanced by climate-related disasters in ethnically fractionalized countries," *Proceedings of the National Academy of Sciences* 113(33), 9216–221.

7. "Sittwe Camp Profiling Report," Danish Refugee Council and United Nations High Comissioner for Refugees, 2017, https:// reliefweb.int/sites/reliefweb.int/files /resources/sittwe_camp_profiling_report _lq.pdf; Jared Ferrie, "Myanmar authorities work to evacuate camps as cyclone nears," Reuters, May 13, 2013.

8. "Central and Northern Rakhine State Case Study: Revisiting emergency response and recovery projects in disaster and conflict affected communities," Food and Agriculture Organization of the United Nations, July 2017, www.fao.org/3/I8564EN /i8564en.pdf. UN Report—Floods in Myanmar Had Devastating Impact on Agriculture," Floodlist, October 22, 2015, http://floodlist.com/asia/un-myanmar -floods-food-security"; "Malnutrition in Myanmar's Rakhine State After Floods," Floodlist, January 26. 2016, http://floodlist .com/asia/malnutrition-spikes-in-myanmars -rakhine-state-after-floods-eu-agency.

9. Both the International Court of Justice and the International Criminal Court have condemned Myanmar's actions and are considering cases against the government and its leaders. "Myanmar Rohingya: What you need to know about the crisis," BBC, January 23, 2020, www.bbc.com/news /world-asia-41566561. Alison Smith and Francesca Basso, "Justice for the Rohingya: What has happened and what comes next?," Coalition for the International Criminal Court, February 13, 2020.

10. "On the Frontlines: Climate Change in Bangladesh," Environmental Justice Foundation, 2018, https://ejfoundation.org// resources/downloads/Climate-Displace-ment-Bangladesh-briefing-2018-v20.pdf.

11. "'It's not a concentration camp': Bangladesh defends plan to house Rohingya on island with armed police," CBC News, February 22, 2018.

12. "2017 Global Report on Internal Displacement," Norwegian Refugee Council, www .internal-displacement.org/global-report /grid2017.

13. For further discussion of this fast-moving field of international law, see Dina Ionesco, "Let's Talk About Climate Migrants, Not Climate Refugees," International Organization on Migration, www.un.org/sustain abledevelopment/blog/2019/06/lets -talk-about-climate-migrants-not-climate -refugees, "UN landmark case for people displaced by climate change," Amnesty International, January 20, 2020, www .amnesty.org/en/latest/news/2020/01 /un-landmark-case-for-people-displaced-by -climate-change; Maya Goodfellow, "How helpful is the term 'climate refugee'?," The Guardian, August 31, 2020, and "Briefing: The concept of 'climate refugee,'" European Parliament, February 2019, www.europarl. europa.eu/RegData/etudes/BRIE/2018 /621893/EPRS_BRI(2018)621893_EN.pdf.

14. More information can be found at Environmental Justice Foundation, www.ejfounda-tion.org; Camillo Boano, Roger Zetter, and Tim Morris, "Environmentally Displaced People: Understanding the linkages between environmental change, livelihoods, and forced migration," Refugees Study Center, University of Oxford, November 2008, https://www.rsc.ox.ac.uk/files/files-1/pb1 -environmentally-displaced-people-2008. pdf, 10; Alexander Betts, ed., "Survival Migration: Failed Governance and the Crisis of Displacement," Journal of Modern African Studies 52(2): 332–33.

15. For a brief summary, see "Global compact for migration," Refugees and Migrants, https://refugeesmigrants.un.org/migration -compact. For the compact's full text, see https://refugeesmigrants.un.org/sites /default/files/180711_final_draft_0.pdf.

16. See more on this case at www.icj-cij.org/en /case/178.

17. See The United Nations and Colonialism, www.un.org/dppa/decolonization/en.

PART III: WARMING WATERS

1. Alanna Mitchell, Sea Sick: The Global Ocean in Crisis (Toronto: McClelland & Stewart, 2009).

2. NOAA estimates that it takes one thousand

years for any given "parcel" of seawater to complete a circuit. https://oceanservice.noaa.gov/education/kits/currents/06 conveyor2.html.

3. IPCC, 2019: IPCC Special Report on the Ocean and Cryosphere in a Changing Climate [H.-O. Pörtner, D.C. Roberts, V. Masson-Delmotte, P. Zhai, M. Tignor, E. Poloczanska, K. Mintenbeck, A. Alegría, M. Nicolai, A. Okem, J. Petzold, B. Rama, N.M. Weyer (eds.)]. In press.

4. Henry, L.G., et al., "North Atlantic ocean circulation and abrupt climate change during the last glaciation," *Science* 353(6298): 470–74.

5. A. Anyamba, J. Chretien, S.C. Britch, et al., "Global Disease Outbreaks Associated with the 2015–2016 El Niño Event," *Scientific Reports* 9(1930), 2019.

6. Bin Wang, Xiao Luo, Young-Min Yang, Weiyi Sun, Mark A. Cane, Wenju Cai, Sang-Wook Yeh, and Jian Liu, "Historical change of El Niño properties sheds light on future changes of extreme El Niño," *PNAS* 116(45), 2019.

7. Luann Dahlman and Rebecca Lindsey, "Climate Change: Ocean Heat Content," www.climate.gov/news-features/understanding-climate/climate-change-ocean-heat-content.

8. Lijing Cheng, John Abraham, Zeke Hausfather, Kevin E. Trenberth. "How Fast Are the Oceans Warming?," *Science*, January 2019, 128–29.

9. For a longer discussion of what has happened in the past when AMOC shuts down, see NASA's Jet Propulsion Lab: http://oceanmotion.org/html/impact/conveyor.htm.

10. Jeff Goodell, "415: The Most Dangerous Number," *Rolling Stone*, May 14, 2019.

11. Ed Yong, *I Contain Multitudes: The Microbes Within Us and a Grander View of Life* (New York: HarperCollins, 2016).

12. All mammals' bodies become warmer when pathogens appear, and many cold-blooded vertebrates seek out warmer environments when they are sick, even if it puts them in greater danger of predation. Plants, which evolved 1.5 billion years ago, are also known to increase their temperature when infected by risky fungi. See Sharon S. Evans, Elizabeth A. Repasky, and Daniel T. Fisher, "Fever and the Thermal Regulation of Immunity: The Immune System Feels the Heat," *Nature Reviews: Immunology* 15(6): 335–49. See also Alan P. Trujillo and Harold V. Thurman, *Essentials of Oceanography*, 12th ed. (Boston: Prentice Hall, 2017).

13. Mitchell, *Sea Sick*, 146–47.

14. Jay Lemery, MD, and Paul Auerbach, MD, *Enviromedics: The Impact of Climate Change on Human Health* (Lanham, MD: Rowman and Littlefield, 2017), 10.

THE ARCTIC OCEAN: WHEN THE ICE MELTS

1. Kyle P. Whyte, "Indigenous Science (Fiction) for the Anthropocene: Ancestral Dystopias and Fantasies of Climate Change Crises," *Environment and Planning E: Nature and Space* 1(1–2): 224–42. More of Whyte's writings on Indigenous climate action can be found at http://kylewhyte.seas.umich.edu.

2. "Skating on Thin Ice: The Thawing Arctic Threatens an Environmental Catastrophe," *The Economist*, April 29, 2017, 16; Chris Mooney, "Greenland's ice losses have septupled and are now in line with its highest sea-level scenario, scientists say," *Washington Post*, December 10, 2019; Arctic Program, "Arctic Report Card," NOAA, 2019; Peter Wadhams, *A Farewell to Ice: A Report from the Arctic* (Oxford, UK: Oxford University Press, 2017), 83; Eric Holthaus, "The Last Time the Arctic Was Ice-Free in the Summer, Modern Humans Didn't Exist," *Slate*, December 2, 2014. Kate Ramsayer, "2020 Arctic Sea Ice Minimum at Second Lowest on Record," NASA, September 21, 2020.

3. For more on this topic, see *Inuit Qaujima-*

jatuqangit: What the Inuit Have Always Known to Be True (Black Point, NS: Fernwood, 2017) 2, 41–59.

4. These children were often taken without the permission or even knowledge of their parents, and the residential schools were often abusive and filthy. See the Nunavut 99 Project, www.nunavut.com/nunavut99/english.

5. For more on the impacts on Indigenous communities, see Igor Krupnik and Dyanna Jolly, eds., *The Earth Is Faster Now: Indigenous Observations of Arctic Environmental Change* (Fairbanks, AK: Arctic Research Consortium of the United States, 2002); Livian Albeck-Ripka, "Why Lost Ice Means Lost Hope for an Inuit Village," *New York Times*, November 25, 2017; "Inuit Qaujimajatuqangit on Climate Change in Nunavut," April 2002, www.gov.nu.ca/environment.

6. Sheila Watt-Cloutier, "It's Time to Listen to the Inuit on Climate Change," *Canadian Geographic*, November 15, 2018.

7. G.K. Healey, *Exploring human health-related indicators of climate change in Nunavut* (Iqaluit, NU: Qaujigiartiit Health Research Centre, March 2015).

8. Ellen Gray, "Unexpected Future Boost of Methane Possible from Arctic Permafrost," https://climate.nasa.gov/news/2785/unexpected-future-boost-of-methane-possible-from-arctic-permafrost. See also the Circumpolar Active Layer Monitoring Network (CALM), a research program of the International Permafrost Association, https://ipa.arcticportal.org/; Intergovernmental Panel on Climate Change, "Special Report on the Cryosphere in a Changing Climate," Part B.1.4, www.ipcc.ch/srocc/chapter/summary-for-policymakers.

9. Methane also persists in the atmosphere for decades more than carbon dioxide. See Carolyn D. Ruppel and John D. Kessler, "The interaction of climate change and methane hydrates," *Reviews of Geophysics* 55(1), March 2017, and Wadhams, *A Farewell to Ice*, 57.

10. Wadhams, *A Farewell to Ice*. Peter Wadhams, "The Global Impacts of Rapidly Disappearing Sea Ice," https://e360.yale.edu/features/as_arctic_ocean_ice_disappears_global_climate_impacts_intensify_wadhams, September 26, 2016.

11. Eric Wolff, "Guest Post: Understanding Climate Feedbacks," Carbon Brief, October 8, 2015; Bobby Magill, "Arctic Methane Emissions 'Certain to Trigger Warming,'" Climate Central, May 1, 2014, www.climatecentral.org/news/arctic-methane-emissions-certain-to-trigger-warning-17374.

12. "Ice-free" is defined as less than one million square kilometers, compared to pre-1980 September averages of six-plus million square kilometers. See "The window to saving Arctic sea ice is rapidly closing," European Space Agency, www.climatechangenews.com/2019/09/18/window-save-arctic-sea-ice-rapidly-closing/, September 19, 2019, and J.A. Screen and C. Desar, "Pacific Ocean Variability Influences the Time of Emergence of a Seasonally Ice-Free Arctic Ocean," *Geophysical Research Letters* 46(4), February 28, 2019.

13. Arrhenius predicted the benefits of longer crop seasons in Scandinavia. Svante Arrhenius, translated by Dr. H. Borns, *Worlds in the Making: The Evolution of the Universe* (New York: Harper and Brothers, 1908).

14. Alexander Sergunin and Valery Konyshev, *Russia in the Arctic: Hard or Soft Power?* (Stuttgart: Ibidem Verlag, 2016), 81–85; Oki Nagai, "China and Russia Battle for North Pole Supremacy," *Nikkei Asian Review*, April 10, 2018; "Status of Arctic Waters Beyond 200 Nautical Miles from Shore," IBRU, Centre for Borders Research, 2016; "Circum-Arctic Resource Appraisal: Estimates of Undiscovered Oil and Gas North of the Arctic Circle," U.S. Geological Survey, 2008, https://pubs.usgs.gov/fs/2008/3049/fs2008-3049.pdf.

15. "Canada's Arctic and Northern Policy Framework," www.rcaanc-cirnac.gc.ca/eng

/1560523306861/1560523330587#s4; A. Atkinson, et al., "Getting It Right in a New Ocean: Bringing Sustainable Blue Economy Principles to the Arctic," WWF Arctic Programme report, November 2018, https://arcticwwf.org/site/assets/files/2050/report_arctic_blue_economy_web.pdf.

16. As of 2018, the largest oil spill in history was 330 million gallons in 1991 in Iraq, followed by the Deepwater Horizon in the Gulf of Mexico in 2010, with 210 million gallons. An oil spill in the Arctic would be nearly impossible to clean up, especially if it occurred in the winter; www.cleanerseas.com/10-worst-oil-spills-in-world-history.

PISCO, PERU: ENSO AND THE END OF FISH

1. Gabriel García Márquez, translated by Edith Grossman, *Love in the Time of Cholera* (New York: Alfred A. Knopf, 1988).

2. Phil Salmon, "ENSO and the Anchovy," Climate Etc., May 11, 2015, judithcurry.com/2015/05/11/enso-and-the-anchovy.

3. A big thank you to Dr. Drew Talley, my friend and colleague who took me through the ENSO process. Any inaccuracies in conveying the science of ocean circulation are my own.

4. For more on this, check out Danielle Torrent Tucker, "Q&A: Tracking the History of El Niño," phys.org, December 20, 2018, phys.org/news/2018-12-qa-tracking-history-el-nio.html#jCp; and ARC Centre of Excellence for Climate Extremes, "'Impossible' research produces 400-year El Niño record, revealing startling changes: Centuries long seasonal record of El Niños from coral cores," *ScienceDaily*, May 6, 2019, www.sciencedaily.com/releases/2019/05/190506111441.htm.

5. From 2005 to 2011, the Peruvian government required parts of their military, schools, and other programs to incorporate anchovies into menus, which did temporarily increase direct consumption of the fish. Once the requirement was lifted, however, programs stopped buying. Read more here: International Fishmeal and Fish Oil (IFFO) Organisation, "Why Feed, Not Food?" www.iffo.net/case-study-peruvian-anchovy-why-feed-not-food.

6. NOAA Fisheries, "Feeds for Aquaculture," FAQ: https://www.fisheries.noaa.gov/insight/feeds-aquaculture.

7. "Perfíles de Pesca y Acuicultura por Países: Perú [Fishing and Aquaculture Profiles by Countries: Peru]," Departamento de Pesca y Acuicultura de la FAO [FAO Fisheries and Aquaculture Department], updated January 15, 2019, www.fao.org/fishery/facp/PER/en.

8. An in-depth look at the different fleets operating within Peru can be found here: José Zenteno, "Peruvian *Anchoveta* Fishery: Industry Structure," *Background Analysis of the Artisanal Sector of the Peruvian* Anchoveta, www.laff.bren.ucsb.edu/sites/default/files/1_Anchoveta%20Industry%20structure.pdf.

9. Malin L. Pinsky, Olaf P. Jensen, Daniel Ricard, and Stephen R. Palumbi, "Unexpected patterns of fisheries collapse in the world's oceans," *PNAS* 108(20): 8317–22, www.pnas.org/content/108/20/8317.full.

10. On a side note, researchers have found that anywhere from 20 percent to 50 percent of the fish sold to consumers at the supermarket and in restaurants is mislabeled. Next time you're thinking of opting for tuna, consider that it takes the average tuna an entire year from the time it dies to when you eat it, often passing through several illegal handling phases that may include human trafficking to food mishandling to organized crime syndicates running the processing plant. SeafoodWatch.org, a scientific nonprofit, has up-to-date information on fisheries practices and the least harmful ways to eat ocean wildlife.

11. The UN's Food and Agriculture Organization maintains a sobering website called

"Illegal, Unreported, and Unregulated (IUU) Fishing" (www.fao.org/iuu-fishing/en/, last accessed on January 20, 2020), and the Pew Foundation's "Ending Illegal Fishing Project" website breaks down the issues in an easy-to-understand way (https://www.pewtrusts.org/en/projects/ending-illegal-fishing-project).

12. "Elinor Ostrom—Biographical," NobelPrize.org, Nobel Media AB 2020, www.nobelprize.org/prizes/economic-sciences/2009/ostrom/biographical.

13. To read more on the commons, see National Research Council, *The Drama of the Commons* (National Academy of Sciences, 2002), www.thecommonsjournal.org, David Bollier's books and blog at www.bollier.org, and the Ostrom Workshop, ostromworkshop.indiana.edu.

14. For an in-depth look at all kinds of crime on the high seas, Ian Urbina's *The Outlaw Ocean* is a thoroughly researched and engrossing read. One of the points that Urbina drives home is that there is simply not enough capacity for oversight or coordinated action.

THE NORTH ATLANTIC: IN DEEP

1. Laffoley and J.M. Baxter, "Explaining Ocean Warming: Causes, scale, effects and consequences," International Union for Conservation of Nature, September 2016, 57.

2. Camilo Mora et al., "How Many Species Are There on Earth and in the Ocean?" *PLoS Biology* 9(8), August 23, 2011.

3. Throughout this chapter are some of Christina's favorite phrases from Rachel Carson's first published essay, "Undersea," which appeared to great acclaim in *The Atlantic* in 1937. Carson had originally written it for the Department of Fish and Wildlife, where she worked as a biologist, but her boss suggested that her writing was too good for a government brochure, and thus launched her career. R.L. Carson, "Undersea," *The Atlantic Monthly*, September 1937.

4. "How Much Oxygen Comes from the Ocean?," National Ocean Service, http://oceanservice.noaa.gov/facts/ocean-oxygen.html, June 12, 2020.

5. Adi Abada and Einat Segev, "Multicellular Features of Phytoplankton," *Frontiers in Marine Science* 5(144), April 24, 2018.

6. For a full explanation of early evolution and endosymbiosis, see Lynn Margulis and Dorion Sagan, *What Is Life?* (Berkeley: University of California Press, 2000). Sagan, an evolutionary biologist, calls this long epoch the Cyanocene in Anna Lowenhaupt Tsing, *The Mushroom at the End of the World: On the Possibility of Life in Capitalist Ruins* (Princeton, NJ: Princeton University Press, 2015).

7. Much later, some of these cyanobacteria-like organisms evolved to become chloroplasts, the parts of a plant cell that allow it to photosynthesize. See Margulis and Sagan, *What Is Life?*

8. Elizabeth Pennisi, "Meet the Obscure Microbe That Influences Climate, Ocean Ecosystems, and Perhaps Even Evolution," *Science*, March 9, 2017; John Waterbury, "Little Things Matter a Lot," *OceanUS*, Woods Hole Oceanographic Institute, March 11, 2005; F. Partensky, W.R. Hess, and D. Vaulot, "Prochlorococcus, a Marine Photosynthetic Prokaryote of Global Significance," *Microbiology and Molecular Biology Reviews* 63(1): 106–27.

9. Carson, *Undersea*.

10. Ibid.

11. Sven Erik Jorgensen and Brian Fath, eds., "Coccolithophore—an Overview," *Encyclopedia of Ecology* (Amsterdam: Elsevier Science, 2008).

12. Shruti Malviya, Eleonora Scalco, Stéphane Audic, Flora Vincent, Alaguraj Veluchamy, Julie Poulain, Patrick Wincker, et al., "Insights into Global Diatom Distribution and Diversity in the World's Ocean," *PNAS* 113(11).

13. This Match-Mismatch Hypothesis, developed by David Cushing in 1969, most severely affects regions of land and sea with high seasonal variability, like the far northern and southern latitudes. Joël M. Durant et al., "Contrasting Effects of Rising Temperatures on Trophic Interactions in Marine Ecosystems," *Scientific Reports* 9(1): 1–9; "Calanus Finmarchicus," COPEPOD—the global plankton database project, NOAA.

14. Veronique LaCapra, "Mission to the Ocean Twilight Zone: The urgent quest to explore one of Earth's hidden frontiers," www.whoi.edu/oceanus/feature/mission-to-the-oceans-twilight-zone.

15. Conversation with Christina Conklin, February 4, 2019.

16. Danielle Hall, "Marine Microbes," Smithsonian Ocean, National Museum of Natural History, July 2019; "Hundreds of Thousands of Viruses Discovered in World's Oceans," *Nature*, April 25, 2019; Corina Brussaard et al., "Marine Viruses," *The Marine Microbiome* (Springer, June 4. 2016); Joshua Weitz and Steven Wilhelm, "An Ocean of Viruses," *Scientist*, July 1, 2013.

17. Designed by a young marine scientist named Alister Hardy, the Continuous Plankton Recorder is towed behind commercial vessels to passively collect plankton at a depth of ten meters, rolling the catch between two thin sheets of silk, which is later analyzed by geographic coordinates. It is the most consistent, long-term data source in ocean science. M. Edwards et al., "Global Marine Ecological Status Report" (Plymouth, UK: Sir Alister Hardy Foundation for Ocean Science, 2016); Andrew D. Barton et al., "Anthropogenic Climate Change Drives Shift and Shuffle in North Atlantic Phytoplankton Communities," *PNAS* 113(11).

18. One of the most notable thinkers is Donna Haraway in *Staying with the Trouble: Making Kin in the Chthulucene* (Durham, NC: Duke University Press, 2016.

19. J.A.N. Lee, "Norbert Wiener," IEEE Computer Society, https://history.computer.org/pioneers/wiener.html.

20. Conversation with Christina Conklin, February 25, 2019. See also Daniel A. Power et al., "What Can Ecosystems Learn? Expanding Evolutionary Ecology with Learning Theory," *Biology Direct* 10(1).

21. More information is available at www.keckfutures.org and www.oceanmemoryproject.com.

22. Ed Yong, "America's New Plan to Zoom In on the Planet's Microbes," *The Atlantic*, May 13, 2016.

23. Conversation between oncological researcher Arun Wiita and Christina Conklin, November 12, 2019.

24. Laffoley and Baxter, "Explaining Ocean Warming," 57.

25. Laffoley and Baxter, "Explaining Ocean Warming," 227.

26. Abigail Tucker, "Jellyfish: The Next King of the Sea," *Smithsonian*, August 2010.

27. In one news story, Japanese fishermen tried cutting up 450-pound Nomura jellies and throwing them back in the ocean, but this caused the jellies to release eggs and sperm, further propagating the problem. See Gaia Vince, "Jellyfish Blooms Creating Oceans of Slime," *BBC Future*, BBC, April 5, 2012.

KISITE, KENYA: CORAL COLLAPSE

1. Ruth Gates in an interview for the documentary film *Chasing Coral*, directed by Jeff Orlowski, Netflix, 2017.

2. In fact, oxybenzone exposure was found to be less of a stressor for some species than rising temperatures. T. Wijgerde, M. Ballegooijen, R. van. Nijland, L. van der Loos, C. Kwadijk, R. Osinga, A. Murk, S. Slijkerman, "Adding insult to injury: Effects of chronic oxybenzone exposure and elevated temperature on two reef-building corals," *Science of the Total Environment* 733 (2020).

3. David Obura and Mishal Gudka, "Western Indian Ocean Post-Bleaching Assessment Training," posted to YouTube by Reef Resilience, July 21, 2017, https://youtu.be /uDBkSbBO-Rk.

4. An accessible explanation of heatwaves is available on NOAA's Research News blog located here: https://research.noaa.gov /article/ArtMID/587/ArticleID/2559 /So-what-are-marine-heat-waves.

5. Information about this coral bleaching event is synthesized from the following two sources: "The Return of the Blob," Kendra Pierre-Louis, *New York Times*, October 21, 2019, and John F. Piatt, et al. "Extreme mortality and reproductive failure of common murres resulting from the northeast Pacific marine heatwave of 2014–2016," *PloS one* vol. 15,1 e0226087, January 15, 2020.

6. A.J. Fordyce, T.D. Ainsworth, S.F. Heron, W. Leggat, "Marine heatwave hotspots in coral reef environments: physical drivers, ecophysiological outcomes and impact upon structural complexity," *Frontiers in Marine Science* 6 (498).

7. Mishal Gudka, David Obura, Jelvas Mwaura, Sean Porter, Saleh Yahya, and Randall Mabwa, "Impact of the 3rd Global Coral Bleaching Event on the Western Indian Ocean i n 2016," Global Coral Reef Monitoring Network (GCRMN)/Indian Ocean Commission, 2018.

8. "Mapping the Global Value and Distribution of Coral Reef Tourism," www.science direct.com/science/article/pii /S0308597X17300635.

9. Ronald Osinga in conversation with Marina Psaros, December 12, 2019.

10. A good read on the ethics of assisted evolution can be found here: K. Filbee-Dexter and A. Smajdor, "Ethics of Assisted Evolution in Marine Conservation," *Frontiers in Marine Science* 6(20).

PINE ISLAND GLACIER: WHAT HAPPENS IN ANTARCTICA DOESN'T STAY IN ANTARCTICA

1. Nicholas Johnson, *Big Dead Place* (Los Angeles: Feral House, 2005), 80.

2. Andrew Shepherd et al., "Mass balance of the Antarctic Ice Sheet from 1992 to 2017," *Nature* 558(219–22).

3. From the European Space Agency blog entry "Emerging Cracks in the Pine Island Glacier," October 18, 2019, www.esa.int/Applications /Observing_the_Earth/Emerging_cracks _in_the_Pine_Island_Glacier.

4. Chris S.M. Turney et al., "Early Last Interglacial ocean warming drove substantial ice mass loss from Antarctica," *Proceedings of the National Academy of Sciences*, February 2020, 117(8) 3996–4006.

5. Robert M. DeConto and David Pollard, "Contribution of Antarctica to past and future sea-level rise," *Nature* 531 (March 2016): 591–97.

6. Two articles that generated a lot of interest for the general public at the time: Eric Holthaus's "Ice Apocalypse," *Grist*, November 21, 2017, and Jeff Goodell's "The Doomsday Glacier," *Rolling Stone*, May 9, 2017.

7. For more, see glaciologist Robert DeCanto's comments in the article, "A Terrifying Sea-Level Prediction Now Looks Far Less Likely," *The Atlantic*, January 4, 2019. Complete findings yet to be published. One of the counter-studies is Tasmin L. Edwards et al., "Revisiting Antarctic ice loss due to marine ice-cliff instability," *Nature* 556 (February 2019): 58–64.

8. Richard Feynman and Jeffrey Phillips-Robbins, *The Pleasure of Finding Things Out: The Best Short Works of Richard P. Feynman* (Cambridge, MA: Perseus Books, 1999).

PART IV:
RISING SEAS

1. Justin Gillis, "Rising Seas Seen as Threat to Coastal U.S.," *New York Times*, March 14, 2012.

2. For more information and interesting case studies, see https://whc.unesco.org/en/climatechange.

3. Zeke Hausfather, "Analysis: How well have climate models projected global warming?," *Carbon Brief*, May 10, 2017.

4. As of May 2020, *Carbon Brief* estimated that reductions in February and March 2020 were approximately 5.5 percent, below the 7.6 percent reductions that the United Nations Environment Program's Emissions Gap Report 2019 projects is necessary to keep warming to 1.5°C.

5. IPCC, 2019: IPCC Special Report on the Ocean and Cryosphere in a Changing Climate [H.-O. Pörtner, D.C. Roberts, V. Masson-Delmotte, P. Zhai, M. Tignor, E. Poloczanska, K. Mintenbeck, A. Alegría, M. Nicolai, A. Okem, J. Petzold, B. Rama, N.M. Weyer (eds.)]. In press.

6. See "Is Sea Level Rising?," NOAA, https://oceanservice.noaa.gov/facts/sealevel.html.

7. The National Oceanic and Atmospheric Administration's Tides and Currents website hosts tide gauge data and maps including the "Sea Level Trends" page: https://tidesandcurrents.noaa.gov/sltrends.

8. Several excellent resources, including mapping tools at www.climatecentral.org and NOAA's Digital Coast Sea Level Rise Viewer provide increasingly accurate projections of where future water will be, but the funding gap remains.

9. As oncologist David Servan-Schreiber writes, "All of us have cancer cells in our bodies. But only some of us will develop cancer." David Servan-Schreiber, *Anticancer: A New Way of Life* (New York: Viking Penguin, 2009).

SHANGHAI, CHINA:
SINK, SANK, SUNK

1. Pan Xiang-chao, "Research on Xi Jinping's Thought of Ecological Civilization and Environment," *Sustainable Development, IOP Conference Series: Earth and Environmental Sciences* 153 (2018): 3.

2. Jonathan Watts, "Shanghai Sinks to New Lows," *The Guardian*, October 6, 2003.

3. This prediction is based on conservative sea-level-rise data, and could be made far worse by up to two meters of subsidence, creating risks far greater than those stated in Jun Wang, Wei Gao, Shiyuan Xu, and Lizhong Yu, "Evaluation of the Combined Risk of Sea Level Rise, Land Subsidence, and Storm Surges on the Coastal Areas of Shanghai, China," *Climatic Change* 115(3–4); Kate Springer, "Soaring to Sinking: How Building Up Is Bringing Shanghai Down," *Time*, May 21, 2012; Xinying Tok, "Sinking Cities: Cracks in the Ground," November 7, 2013, www.chinawaterrisk.org/opinions/sinking-cities-cracks-in-the-ground.

4. Shui-Long Shen and Ye-Shuang Xu, "Numerical Evaluation of Land Subsidence Induced by Groundwater Pumping in Shanghai," *Canadian Geotechnical Journal* 48(9).

5. Yang Jian, "Metro Blamed for Subsidence, Cracks in Residential Building," *Shanghai Daily*, April 10, 2014, https://archive.shine.cn/metro/society/Metro-blamed-for-subsidence-cracks-in-residential-building/shdaily.shtml.

6. "Shanghai Plans Dam on Huangpu River," China Through a Lens, *China Daily*, February 9, 2004, www.china.org.cn/english/2004/Feb/86647.htm; Josh Holder, Niko Kommenda, and Jonathan Watts, "The Three-Degree World: Cities That Will Be Drowned by Global Warming," *The Guardian*, November 3, 2017; S.F. Balica, N.G. Wright, and F. van der Meulen, "A Flood Vulnerability Index for Coastal Cities

and Its Use in Assessing Climate Change Impacts," *Natural Hazards* 64(1).

7. "China Is Trying to Turn Itself into a Country of 19 Super-Regions," *The Economist*, June 23, 2018.

8. The most conservative models (together called RCP 2.6) assume we never surpass an atmospheric carbon load of five hundred parts per million (ppm), though we are currently on course to reach five hundred ppm by 2050. Nicola Jones, "How the World Passed a Carbon Threshold and Why It Matters," *Yale Environment 360*, January 26, 2017, https://e360.yale.edu/features/how-the-world-passed-a-carbon-threshold-400ppm-and-why-it-matters; John A. Church and Peter U. Clark, coordinating lead authors, "Sea Level Change," Intergovernmental Panel on Climate Change (Cambridge, UK: Cambridge University Press, 2013), 1186–87.

9. "China's Pioneering City of Dongtan Stalls," *The Telegraph*, October 18, 2008.

10. Julie Sze, *Fantasy Islands: Chinese Dreams and Ecological Fears in an Age of Climate Crisis* (Berkeley: University of California Press, 2015), 36.

11. Ji Jing, "Mountains of Gold and Silver," *Beijing Review*, April 10, 2020.

12. For more information on ecological civilization, see www.processcenturypress.com/publications/series/toward-ecological-civilization-series. See also Dimitri de Boer, "Will China's New Ministerial Structure Help the Environment?," www.clientearth.org/will-new-chinese-ministerial-structure-help-environment, March 21, 2018.

13. Linli Cui and Jun Shi, "Urbanization and Its Environmental Effects in Shanghai, China," *Urban Climate* 2 (December 1, 2012).

14. The British Ecological Society, "The ecosystem: An evolving concept viewed historically," *Functional Ecology* 11 (1997): 268–71.

15. U.S. Climate Resilience Toolkit, https://toolkit.climate.gov/topics/ecosystems.

16. Arne Naess, translated and edited by David Rotherberg, *Ecology, Communitiy and Lifestyle: Outlines of an Ecosophy* (Cambridge, UK: Cambridge University Press, 1989).

17. Yang Jian, "Land subsidence reduced to safe level," archive.shine.cn/metro/public-services/Land-subsidence-reduced-to-safe-level/shdaily.shtml, April 23, 2016; Jun Wang et al., "Evaluation of the Combined Risk of Sea Level Rise," *Climatic Change* 115(3–4).

18. Jun Wang, Shiyuan Xu, Mingwu Ye, and Jing Huang, "Overtopping Risk Assessment of Seawalls and Levees in Shanghai," *International Journal of Disaster Risk Science* 2(4): 32–42.

19. Jun Wang, et al., "Evaluation of the Combined Risk of Sea Level Rise."

20. Steven Lee Meyers, "China's Pledge to Be Carbon Neutral by 2060: What It Means," *New York Times*, September 25, 2020.

21. Such products are currently being developed: www.sustainablebuild.co.uk/environmentally-friendly-concrete.html and www.bbc.com/news/science-environment-46455844.

22. According to the Council on Foreign Relations, the Belt and Road Initiative has led to investment and construction projects backed by Chinese funding, expertise, influence, and labor in sixty countries to date: "China's overall ambition for the BRI is staggering," www.cfr.org/backgrounder/chinas-massive-belt-and-road-initiative.

23. Finn Aberdein, "Chairman Zhang's Flat-pack Skyscrapers," BBC, www.bbc.co.uk, June 11, 2015.

24. The IPCC projects a range for sea-level rise in 2050 of 0.4 to 0.7 meters (1 to 2 feet). "Sea Level Change," IPCC, 1204.

HAMPTON ROADS, VIRGINIA: BYE, BYE, BIRDIES

1. Jane Goodall, *Seeds of Hope: Wisdom and Wonder from the World of Plants* (New York: Grand Central Publishing, 2014).

2. The Chesapeake Bay Program, a public/nonprofit initiative to monitor and restore the bay, hosts an informative website with accessible summaries of research, education, and monitoring taking place around the bay. A field guide to the birds in the region can be found at: https://www.chesapeakebay.net/discover/field-guide/all/birds/all.

3. Elizabeth Rush, *Rising: Dispatches from the New American Shore* (Minneapolis: Milkweed Editions, 2018).

4. Jennifer Howard et al., "Clarifying the role of coastal and marine systems in climate mitigation," *Frontiers in Ecology and the Environment,* February 1, 2017.

5. One of the challenges with eastern black rail conservation is that relatively little is known about them. One of the most comprehensive reviews of data is found in Bryan Watts's 2016 report "Status and distribution of the eastern black rail along the Atlantic and Gulf Coasts of North America," Center for Conservation Biology Technical Report Series, CCBTR-16-09, College of William and Mary, Williamsburg, Virginia.

6. The Center for Biological Diversity filed a lawsuit in March 2020, and the Fish and Wildlife Service has completed initial research and a recommendation to list the species: www.fws.gov/southeast/faq/proposed-listing-for-the-eastern-black-rail.

7. "Tidal Wetlands Protection in Virginia: Time for an Update," report for the College of William and Mary and Virginia Sea Grant, Spring 2014.

8. IPBES, "Summary for policymakers of the global assessment report on biodiversity and ecosystem services of the Intergovernmental Science-Policy Platform on Biodiversity and Ecosystem Services," 2019, ipbes.net/system/tdf/inline/files/ipbes_global_assessment_report_summary_for_policymakers.pdf?file=1&type=node&id=36213.

9. Rick Hanson, *Resilient: How to Grow an Unshakeable Core of Calm, Strength, and Happiness* (New York: Harmony Press, 2018).

10. Christina Conklin in conversation with Nate Hagens, June 6, 2019. See also www.postcarboninstitute.org. Ellie Cohen in conversation with Marina Psaros, June 2020.

11. The 2018 documentary film "Rodents of Unusual Size" tracks how Louisiana residents have attempted to manage their own runaway nutria populations, www.rodentsofunusualsize.tv.

BEN TRE, VIETNAM: DOING MORE WITH LESS

1. "Rice Is Life—Vietnam," International Year of Rice 2004, www.fao.org, 2004.

2. Nguyen Duc and his family are a fictional representation of rice-farming families compiled from numerous videos and articles.

3. Mark Scialla, "Meet the Mekong Delta Rice Farmers Who Are on the Frontline of Sea Level Rise," *Vice*, May 13, 2015.

4. "Local Demonstration Projects on Climate Change Adaptation," Mekong River Commission, October 2014, www.mrcmekong.org/assets/Publications/Reports/Local-demonstration-projects-on-CCA-final-report-of-1st-batch-project-in-Vietnam.pdf, 22–23.

5. "Bến Tre rice farmers switch to other crops," *Viet Nam News*, July 21, 2018.

6. This is a composite quote from a *Guardian* story about several farmers. Kit Gillet, "Vietnam's rice bowl threatened by rising seas," August 21, 2011.

7. "Bến Tre rice farmers switch to other crops," *Viet Nam News*.

8. Mekong River Commission for Sustainable Development, "Stories from the Mekong, Living on the edge of the rising sea," www.mrcmekong.org.

9. World Bank Group, "Transforming Vietnamese Agriculture: Gaining More from Less," Vietnam Development Report 2016.

10. Mekong River Commission, "Stories from the Mekong."
11. "Government Resolution 120/NQ-CP on Sustainable and Climate-Resilient Development of the Mekong Delta of Viet Nam," www.mekongdelta.com, November 17, 2017.
12. Alex Chapman and Van Pham Dang Tri, "Climate change is triggering a migrant crisis in Vietnam," www.phys.org, January 9, 2018.
13. Alex Smajgl, "Climate Change Adaptation Planning in Vietnam's Mekong Delta," Mekong Region Futures Institute (MERFI), November 2018, 2.
14. Liz Koslov, "The Case for Retreat," *Public Culture* 28(2[79]): 359–87, May 1, 2016.
15. I dislike the term *sustainable* and have rarely used it in this book, but this is one of the few scenarios—retreat from rising seas—that could be an opportunity to choose sustainable, long-term solutions that do not merely repackage the status quo.
16. The Dutch have long promoted themselves as the world's leading experts on flood management. When the country shifted its international development policy from "aid to trade" in 2008, the government polled water experts, who advised that Vietnam would be the best partner to approach with Dutch expertise.
17. Margreet Zwarteveen et al., "Dutch delta masterplans: What makes them travel and what makes them Dutch?," FLOWs—The Water Governance Blog at the Delft Institute for Water Education, June 29, 2018.
18. "Government Resolution 120/NQ-CP."
19. M. Van Dijk et al., "Land Use, Food Security, and Climate Change in Vietnam," *Wageningen University & Research Policy Brief*, September 2012, 9.
20. Seed-to-Table, "Ecologically Integrated Farming System in the South of Vietnam: Rice-Duck-Shrimp Farming in Ben Tre Province," www.fao.org/3/a-be863e.p.

THE THAMES ESTUARY, BRITAIN: FROM GRAVESEND TO ALLHALLOWS

1. George Monbiot, www.monbiot.com/about.
2. Sean McPolin, "Gravesham council declare climate change emergency as Extinction Rebellion youngster questions council," *Kent On-line*, June 26, 2019.
3. Ibid.
4. Correct as of June 2020, www.gov.uk/guidance/the-thames-barrier.
5. *Thames Estuary 2100: Managing flood risk through London and the Thames Estuary,* UK Environment Agency, November 2012, 134–35.
6. Caroline Crampton journeys by sail along the Thames, excavating layers of history and literature along the way in *The Way to the Sea* (London: Granta, 2019).
7. *Thames Estuary 2100,* 28.
8. The initial concept for Thames 2100 was to forecast the rest of the century, but planners and officials objected that they couldn't accurately forecast that far into the future, so they chose a nearer-term timeline instead. *TE2100 5 Year Review: Non-Technical Summary,* July 2016, https://assets.publishing.service.gov.uk/government/uploads/system/uploads/attachment_data/file/558631/TE2100_5_Year_Review_Non_Technical_Summary.pdf and Thames Estuary 2100.
9. Most of the region has been designated P3, that is, maintaining the current defenses, with some areas P4 and P5, i.e. needing some measure of additional flood protection. Almost no land has been rated P1, or land that will not be defended against future floods. *Thames Estuary 2100.* See also "Vision 2050," *Thames Estuary 2050 Growth Commission,* June 2018, 2–8, https://assets.publishing.service.gov.uk/government/uploads/system/uploads/attachment_data/file/718805/2050_Vision.pdf, and *The Draft New London Plan,* www.london.gov.uk/sites/default/files/draft_london_plan

_-_consolidated_changes_version_-_clean
_july_2019.pdf.

10. Conversation with Christina Conklin.

11. The first town to be slated for abandonment is 2045 in North Wales, because it lies on a spit of land between lagoon and sea and has been deemed not viable once seas begin to rise more rapidly. Tom Wall, "'This is a wake-up call': The villagers who could be Britain's first climate refugees," *The Guardian Weekend*, 24–32.

12. One such recently discovered site of transilience is the archaeological site in South Dakota where Robert DePalma identified rocks formed on the day of the earth-shattering asteroid that killed the dinosaurs 66 million years ago. William J. Broad and Kenneth Chang, "Fossil Site Reveals Day That Meteor Hit Earth and, Maybe, Wiped Out Dinosaurs," *New York Times*, March 29, 2019.

13. The best resource for accurate climate science is "Managing the Coast in a Changing Climate," Committee on Climate Change, October 26, 2018.

14. "Vision 2050," *Thames Estuary 2050 Growth Commission*, 9.

15. In addition, Scotland is rising by a centimeter every twenty years or so as a result of glacial retreat thousands of years ago. As a result, southern England is naturally subsiding by about the same amount. "New coastland map could help strengthen sea defences," *Durham University News,* www.dur.ac.uk, October 7, 2007.

16. The Asian tiger mosquito is already present in Italy and a few other southern European countries, where young children have no acquired immunity. World Health Organization, www.who.int/malaria/areas /high_risk_groups/children/en/, www.researchitaly.it/en/projects/tiger-mosquito-italy-to-lead-first-european-project-to-respond-to-the-emergency.

17. Bruno Latour, *Down to Earth: Politics in the New Climate Regime* (London: Polity Books, 2018).

ISE, JAPAN: TRADITION FOR THE FUTURE

1. Excerpted from the speech by Martin Palmer, a longtime advisor to the United Kingdom's Prince Philip, who founded the Alliance of Religions and Conservation in 1995, at the launch of FaithInvest.org, October 31, 2018. See www.faithinvest.org/news and www.arcworld.org.

2. Paul Vallely, "History in the making: An unprecedented visit to Ise Jingu, Japan's holiest shrine, to see it rebuilt under the beliefs of the Shinto religion," *Independent*, June 22, 2014.

3. Poor forest management had meant these logs had to be imported in recent centuries, but starting in the 1920s, Shinto leaders began stewarding their forests better, and this time around, some smaller trees could be taken from Ise-Jingu's sacred forest for the first time since 1391. Vallely, "History in the making."

4. Paul Vallely, "The day an ancient taboo bowed out," *Independent*, June 2, 2014, Alliance of Religions and Conservation, www.arcworld.org.

5. For details see https://fore.yale.edu/ Climate-Emergency/Climate-Change -Statements-from-World-Religions and Mary Evelyn Tucker and John Grim, "Introduction: The Emerging Alliance of Religions and Ecology," *Daedelus*, Fall 2001.

6. For a detailed review of this storm, as well as several other historic floods and disasters, see "1959 Super Typhoon Vera: 50-Year Retrospective," Risk Management Solutions, Inc., 2009, www.rms.com /publications/natural-catastrophes. See Climate Central's Surging Seas Risk Zone Map for current and future sea levels around the world, https://sealevel.climatecentral.org.

7. Erik Kirschbaum, "Seas May Rise 2.3 Meters per Degree C of Global Warming: Report," *Scientific American*, July 15, 2013.

8. In Japanese cities, virtually every green space is a shrine, with meticulously pruned trees and swept sidewalks, and every village has a small shrine nearby, hung with ceremonial ropes and paper chains in various states of decay. See www.bbc.co.uk/religion/religions/shinto.shtml, September 16, 2009.

9. Umair Irfan, "Japan is king of efficiency. But it's losing climate passion," *Environment & Energy Publishing*, October 3, 2017, www.eenews.net.

10. Ryusei Takahashi, "Japanese activists join global climate strike ahead of U.N. summit on global warming," www.japantimes.co.jp, September 20, 2019.

11. Tatiana Schlossberg, "Japan Is Obsessed with Climate Change. Young People Don't Get It," *New York Times*, December 5, 2016.

12. Michel Serres, translated by Randolph Burks, *Biogea* (Minneapolis: Univocal, 2012), 129.

13. "Laudato Si: On Care for Our Common Home," www.vatican.va/content/francesco/en/encyclicals/documents/papa-francesco_20150524_enciclica-laudato-si.html.

14. Notably, he cites Michel Serres, the noted French philosopher and mathematician, whose many works consider the relation of nature and culture, especially *The Five Senses* and *Biogea*. T.J. Demos, *Decolonizing Nature: Contemporary Art and the Politics of Ecology* (Berlin: Sterberg Press, 2016), 181–203.

15. Jill Baker, "From Zero to Hero: How Japan Inc came from behind to lead on climate risk reporting," www.ethicalcorp.com/zero-hero-how-japan-inc-came-behind-lead-climate-risk-reporting; Climate Disclosure Project, www.cdp.net/en, and http://there100.org/companies.

16. Matthew Green, "Make climate fight 'sexy,' says Japan's new environment minister," *Reuters*, September 22, 2019.

17. Hiroko Tabuchi, "Japan Races to Build New Coal-Burning Power Plants, Despite the Climate Risks," *New York Times*, February 5, 2020.

18. Renee Juliene Karunungan, "Coal power plants in Bataan commit human rights violations," *Rappler*, August 5, 2015, www.rappler.com/corruption/101632-bataan-coal-plants-commit-human-rights-violations; Market Forces Letter to Gregory C. Case, President & CEO of AON Insurance, www.marketforces.org.au/wp-content/uploads/2019/07/Aon-Van-Phong-1-letter-27Jun2019.pdf.

19. Yumiko Murakami, "Climate change: A race against time," *Japan Times Opinion*, September 19, 2018, www.japantimes.co.jp.

20. "Commitment N° 11—Commitment by Sovereign Funds," One Planet Summit: A Platform of Commitments to Meet the Challenge of Climate Change, July 6, 2018, www.oneplanetsummit.fr/en/commitments-15/commitment-sovereign-funds-35.

21. Alliance of Religions & Conservation, "International Religious Forestry Standard," 2014, www.arcworld.org/projects.asp?projectID=344.

22. Victoria Finlay et al., "The ZUG Guidelines to faith-consistent investing," Alliance of Religions and Conservation, October 20, 2017, 5–6; "Major new alliance of religious investment funds creating a better and fairer world," Alliance of Religions and Conservation, Press Release, November 2, 2017, www.arcworld.org.

23. Vallely, "History in the making."

TOWARD TRANSILIENCE

1. "Let This Darkness Be a Bell Tower," *On Being*, December 8, 2016.

2. Works that have had a major impact on my thinking include James Lovelock, *A Rough Ride to the Future* (New York: Penguin, 2015) and *The Revenge of Gaia* (New York: Basic Books, 2007). Influential works include Gilles Deleuze, *Spinoza: Practical Philosophy* (San Francisco: City Lights, 2001); Bruno Latour, translated by Catherine

Porter, *Facing Gaia: Eight Lectures on the New Climatic Regime* (Cambridge, UK: Polity, 2017) and *Down to Earth: Politics in the New Climatic Regime* (Cambridge, UK: Polity, 2018); Michel Serres, translated by Margaret Sankey and Peter Cowley, *The Five Senses: A Philosophy of Mingled Bodies* (London: Bloomsbury Academic, 2016) and *Biogea*, translated by Randolph Burks (Minneapolis: Univocal, 2012); Pieter Sloterdijk, translated by Wieland Hoban, *Foams: Spheres*, Vol. III (Cambridge, MA: Semiotext(e), 2016); Donna Haraway, *Staying with the Trouble: Making Kin in the Chthulucene* (Durham, NC: Duke University Press, 2016).

3. Thomas Kuhn, the philosopher of science, first defined a paradigm shift as those rare pivot points in scientific understanding when new information makes an entire commonly held worldview obsolete. The process of acceptance is slow at first because it is always easier to maintain the status quo. But some people do make the leap across the chasm, and others follow, until a trickle becomes a stream becomes a flood. Thomas Kuhn, *The Structure of Scientific Revolutions* (Chicago: University of Chicago Press, 1962).

4. Ralph Ranalli, "Erica Chenowith Illuminates the Value of Nonviolent Resistance in Societal Conflicts," www.hks.harvard.edu/faculty-research/policy-topics/advocacy-social-movements/paths-resistance-erica-chenoweths-research.

5. Craig Macartney, "Christians, Called to Steward Creation," *Christian Week*, December 4, 2014.

6. This is a phrase used by the climate change leader Joanna Macy in her program *The Work That Reconnects*, www.joannamacy.net.

7. This is already happening in some parts of Africa and Asia, even as China and Japan build cheap, dirty coal-fired power plants in others.

8. Such a quantifiable plan exists in Paul Hawken, ed., *Drawdown: The Most Comprehensive Plan Ever Proposed to Reverse Global Warming* (New York: Penguin, 2017), updated regularly at www.drawdown.org.

IMAGE SOURCES

xvi–xvii Bojan Savric, David Burrows, and Melita Kennedy, "The Spilhaus World Ocean Map in a Square," storymaps.arcGIS.com

3 Lokrantz/Azote, based on Steffen et al. 2015

5 Gruber 2011

6 Jambeck et al. 2015, Lebreton et al. 2018, PBS *NewsHour*, GoogleEarth

9 United Nations Environment Program, GRID-Arundel

11 National Oceanographic and Atmospheric Administration (NOAA)

13 Theoceancleanup.com based on Lebreton et al. 2017

15 Breitburg et al. 2018, *Yale Environment 360*, Goes/Gomes Lab at Columbia University, National Aeronautics and Atmospheric Administration (NASA), Google Earth

16 Breitburg et al. 2018

20 International Union for the Conservation of Nature

22 NOAA

25 International Panel on Climate Change

26 Food and Agriculture Organization of the United Nations, 2011, "Food Wastage Footprint & Climate Change," www.fao.org/3/bb144e/bb144e.pdf (Reproduced with permission)

29 Macormack/Cook Island News, Government of the Cook Islands, Google Earth

30 Coral Digest

34 San Francisco Estuary Institute, Bay Area Water Boards, Baykeeper, Google Earth

36 California Office of Environmental Hazard Assessment

38 Agency for Toxic Substances and Disease Registry

40 Matthew Ballew, Jennifer Marlon, Seth Rosenthal, Abel Gustafson, John Kotcher, Edward Maibach, and Anthony Leiserowitz, "Do younger generations care more about global warming?" Yale Program on Climate Change Communication, June 11, 2019

43 Velden et al. 2017 and Rohde/realclimate.org

47 *New York Times*, Texas Commission on Environmental Quality, *Climate Central*, Center for Land Use Interpretation, Google Earth

48   NOAA

51   U.S. Army Corps of Engineers, National Research Council

52   U.S. Environmental Protection Agency, www.epa.gov/americaschildrenenvironment/ace-environments-and-contaminants-contaminated-lands

54   *Climate Central*, Google Earth

57   World Meteorological Organization, 2009, "Selecting Measures and Designing Strategies for Integrated Flood Management," WMO-No. 1047, www.floodmanagement.info/guidance-document/

61   *Climate Central*, Google Earth

63   NOAA

66   C40.org

68   NASA, NOAA, *Washington Post*, Google Earth

71   National Hurricane Center

74   National Public Radio, United Nations Office for the Coordination of Humanitarian Affairs, Google Earth

77   Kumari Rigaud, Kanta, Alex de Sherbinin, Bryan Jones, Jonas Bergmann, Viviane Clement, Kayly Ober, Jacob Schewe, Susana Adamo, Brent McCusker, Silke Heuser, and Amelia Midgley, 2018, *Groundswell: Preparing for Internal Climate Migration* (Washington, DC: The World Bank)

78   Environmental Migration Portal

82-3   NOAA

87   Rahmsdorf 2002

91   Peter Wadhams, National Snow and Ice Data Center (NSIDC), Google Earth

93   NSIDC

94   NSIDC

96   Pelaudeix/Arctic Portal

98   Takahashi/NOAA, Google Earth

102   Allison et al. 2009

107   NOAA, Google Earth

109   Brotz et al. 2012, *Yale Environment 360*

115   Marine World Heritage, *The New Republic*, Google Earth

119   NOAA

120   W.J. Skirving, S.F. Heron, B.L. Marsh, G. Liu, J.L. De La Cour, E.F. Geiger, and C.M. Eakin (2019) "The relentless march of mass coral bleaching: a global perspective of changing heat stress," *Coral Reefs* 38, 547, DOI:10.1007/s00338-019-01799-4

123   Rignot et al. 2011, Commonwealth of Australia, antarcticglaciers.org

126   NASA's Goddard Space Flight Center

131   NASA's Goddard Space Flight Center

132   NASA

134   *Climate Central*, newgeography.com, Google Earth

136   chinawaterrisk.org

137   John Lambert (pre-publication data), *Global Facility for Disaster Reduction and Recovery*

139   PBL Netherlands Environmental Assessment Agency

144   Long Island Sound Study/Graphic by Lucy Reading-Ikkanda, http://longislandsoundstudy.net/wp-content/uploads/2015/08/SLAMMdid-you-know-fact-sheet2-V05.pdf

147   Kenneth V. Rosenberg, Adriaan M. Dokter, Peter J. Blancher, John R. Sauer, Adam C. Smith, Paul A. Smith, Jessica C. Stanton, Arvind Panjabi, Laura Helft, Michael Parr, and Peter P. Marra, "Decline of the North American Avifauna," *Science,* October 4, 2019

148   D'Odorico 2012

150   FAO, *New York Times*, *Climate Central*, Google Earth

152   World Resources Institute

155   Chi-Chung Chen, Bruce McCarl, and Ching-Cheng Chang, "Climate change, sea level rise and rice: Global market implications," *Climatic Change* 110, 543–560 (2012), https://doi.org/10.1007/s10584-011-0074-0

159   *Climate Central*, Google Earth

160   Environment Agency, 2019, "Exploratory Sea Level Projections for U.K. to 2300"

164   *Climate Central*

166   United National Educational, Scientific, and Cultural Organization (UNESCO), jhti.berkeley.edu, Marzeion and Levermann/Environmental Research Letters 2014, Google Earth

169   U.S. Energy Information Administration, www.cia.gov/international/analysis/country/JPN

170   Marzeion and Levermann 2014

INDEX

Page numbers in italics represent maps and illustrations.

acid rain, 4, 5

adaptation. *See* climate adaptation

adaptation engineering, 58–59

addiction and recovery, science of, 3–4

agricultural productivity: percentage impacts of sea-level rise on agricultural land, *155*; projected (year 2050), *152*

albedo, 93, *94*

Albert, Gerrard, 186n11

algae, xii, 1–2, 106–9; blooms, 2, 18, 20, *31*, 33, 87, 109, 110; large-scale farms, 19–20; toxic algal blooms, 2, *31*, 33. *See also* plankton

Alliance of Religions and Conservation, 167, 172, 204n1

Alphas (children of Millennials), 38–40

American Association of Retired People, 11

American Medical Association, 11

Amundsen Sea, 128

Anam, Tahmima, 75

anchovy fishing industry (Peru), 99–105

Antarctic Treaty, 125–26, 127–28

Antarctica: Antarctic Treaty and signatories, 125–26, 127–28; ecotourism, 127, 128; glacial retreat, 122–28; land ice loss and sea-level rise, 122–28, *126*; melting of Pine Island Glacier, 122–28, *123*; West Antarctic Ice Sheet, 124–25, 127

Arabian Sea, 14–21, *15*; food web and *Noctiluca* blooms, 14–17, *15*, *17*; overfishing and species extinction, 17, 18, 19–20; oxygen minimum zone (OMZ), 14–19, *15*, 21; regime shift, *16*, 18–19

archaea, 110, *111*

Arctic Ocean warming, 90–97, *91*; albedo, 93, *94*; first-year sea ice, 90, *91*, *93*; frazil ("grease ice"), 90; and global climate regulation, 90–95, *91*, *94*; ice-albedo feedback and Arctic amplification, 93, *94*; Inuit communities, 92–93, 95–97; multi-year sea ice, 90, *91*, *93*; oil/gas mining and drilling, 95, 96–97; pancake ice, 90; permafrost melting, 93–94, 96; positive feedback loops, 93–94; sea ice melting, 90–97, *91*, *93*; shipping and trade routes, 95, *96*

Arctic Policy Framework (Canada), 95

Argentina, 127

Arkema chemical plant (Crosby, Texas), 49

Army Corps of Engineers, 63, 70, 190n11

Arrhenius, Svante, 95

artificial intelligence (AI), 104, 112

Asian tiger mosquito (*Aedes albopictus*), 163–64, 204n16

assisted evolution, 117–18

Association of Shinto Shrines, 167

Atlantic Meridional Overturning Circulation (AMOC), 86

atmospheric aerosol loading, *3*, 10

Auerbach, Paul, 89

Aung San Suu Kyi, 79

Australian Institute of Marine Science, 118

Bangkok, Thailand, *164*

Bangladesh: Kutupalong Camp in Cox's Bazar, *74*, 75, 77–78, 80; Rohingya refugees and border with Myanmar, *74*, 75, 77–79, 81; sea-level rise projections, *164*; Sundarbans, 130, *164*; Thengar Char refugee camp, 77–78

Bay of Bengal, *74*, 76

Baykeeper, 35–36, 37

Bến Tre, Vietnam, *150*, 153, 156–57. *See also* Mekong Delta of Vietnam

benzene, 183n12

Bering Strait, *xvii*

Berry, Thomas, 32

Bibi, Setera, 75, 77

biogeochemical feedback loops, 1, *3*, 10, 18, 89, 106, 181n2

biological pump, 108–9

biosphere integrity, *3*, 10

bird populations: Chesapeake Bay's estuarine wetlands, 142–49; eastern black rail, 145–47, 149; North America (change since 1970), *147*; seabirds, 7, 102, 117; specialist species, 145–46

bisphenol-A, 9

"The Blob" (marine heatwave of 2013–2015), 116–17

Bloomberg, Michael, 60

body, human. *See* health and human bodies

body, the ocean as, xiii–xiv, 88–89, 178

Bohm, David, 170

bonds, government, 50

BP, 65

Breitberg, Denise, 18

bristlemouth, 110

bromophenols, 1

Brown, Mark, 31, 32

Buddhists: Japanese, 168; Myanmar, 76–77, 79

Bullard, Robert, 49–50

Calanus finmarchicus (copepod), 109–10

calcium carbonate, 24, 108, 116

California: plastics regulation, 11; San Francisco Bay Area toxics pollution, *34*, 34–40

Camden, Maine, *22*, 23–25

Canadian Arctic, *91*, 92–93, 95–97. *See also* Arctic Ocean warming

cancer-causing chemicals, 9, 10, 183n12. *See also* PFAS (per- and polyfluoroalkyl substances)

cap and trade programs, 5

capitalism, 4, 35, 103

carbon dioxide: atmospheric levels, 88, 89, 96, 138, 201n8; carbon sequestration, 2, 20, 88, 93–94, 108–9, 141, 144; cement-related emissions, *139*; ocean absorption, 2; and ocean acidification, 24. *See also* greenhouse gas emissions (GHG)

carbon sequestration, 2, 20, 88, 93–94, 108–9, 141, 144

carbon taxes, 27, 65

carbonic acid, 2, 24

Carson, Rachel, 197n3

"cat bonds" (catastrophe bonds), 50

cement production: cement-related emissions, *139*; eco-cements, 141

Center for Conservation Biology, 145

Center for Disease Control, 36

Chen Jining, 135

Chesapeake Bay Program, 202n2

Chesapeake Bay's estuarine wetlands, 142–49, *143*; bird populations, 142–49; the eastern black rail, 145–47, 149; sea-level rise, 142–49; and suburban development pressures, 145

Chevron, 65

chikungunya, 164

China, 80, 135–41; and Antarctic Treaty, 125; Arctic shipping and trade routes, 95, *96*; Belt and Road Initiative, 201n22; food waste, 25; government-backed environmentally sustainable projects, 138, 140–41; Pearl River Delta sea-level rise projections, *164*; plastics production and pollution, 8, 12, 182n6, 183n20; polyester manufacturing, 8, 182n6; water security, *136*; Yangtze River region flooding and land subsidence, *134*, 135–41. *See also* Shanghai, China

chlorofluorocarbons (CFCs), 4

chloroplasts, 197n7

ciguatera poisoning, 28–30

Circumpolar Deep Water Current, 124

citizens, what we can do as, 174–75

Clean Air Act Amendment (1990), 4–5

Clean Water Act, 145

climate adaptation, ix–x, 26; adaptation engineering, 58–59; balancing economic development and, 154–57; deep, xii; definitions, 26; "hard" options, 26; and sea-level rise, 60–67, 144; "soft" options, 26; uncertainties, x, 126. *See also* climate mitigation

climate migrants, *78*, 79, 80–81, 153–54

climate mitigation, ix, 26, 144, 145. *See also* climate adaptation

climate refugees, 79

climate risk management, x, 50; contingent decisions, x; flood risk management, *51*, *57*; and government bonds, 50; scenario-planning, x

clothing production: fast fashion, 12; and plastics, 7–9, 12–13; renewable fibers, 12–13, 184n22

Cloutier, Sheila Watt, 92

coal-burning power plants in Japan, *169*, 171

coastal cities: and sea-level rise, *34*, *134*, *137*, *143*, 150, 159, *164*, 166; and strengthening storms, 42–43, 60, 66, 140. *See also* sea-level rise and coastal cities

Coastal City Flood Vulnerability Index, 137

coastal wetlands. *See* estuarine wetlands

coccolithophores, *108*, 108–9

coccoliths, 108–9

colonialism. *See* decolonization

Columbia University, 14, 60, 62

Committee on Climate Change (CCC) (UK), 163

Committee on Climate Change (Pakistan), 21

commons (definitions), 103

community dialogues about climate change, 175

Comoro archipelago, 80–81, *115*

ConocoPhillips, 65

consumers, what we can do as, 174

Continuous Plankton Recorder (CPR), 111, 198n17

Convention on International Trade in Endangered Species (CITES), 105

Cook Islands, 28–33, *29*; development-free zones and Marae Moana protected area, *29*, 30–32; manganese nodules and seabed mining, *29*, 30–33

Cook Islands Investment Corporation (CIIC), 32–33

copepods, *108*, 109–10

copolymers, 11

corals, 24, 30, 114–21, *115*; approaches to protection and restoration, 117–18; bleaching events, 89, 116–17, *120*; coral farm cultivation and domestication, 117, 118–20; coral reef tourism, 117, 118–21; and *Gambierdiscus* protist, 30; Kenya, 114–21; as keystone species, 114; map of coral stress, *119*; ocean warming and coral reef collapse, 114–21, *115*; polyps, 114–16, *116*; reef ecosystems, 114; sanctuaries, 120–21

COVID-19 pandemic, xi, 64–65, 130, 163, 164, 174, 176–77

Cox's Bazar, Bangladesh, *74*, 77–78, 80

"cradle-to-cradle" models of manufacturing, 11–12

ctenophores (sea gooseberries), 110

cyanobacteria, 106–8, 197n7

cybernetic theory, 111

Cyclone Herwart (2017), 56

Cyclone Komen (2015), 77

cyclones, tropical (hurricanes or typhoons), 42, *43*, 46–48, *68*, *71*, 140; Caribbean, *68*; Cyclone Herwart in Europe, 56; Cyclone Komen and Myanmar, 77; Hurricane Harvey and Houston, 42, 46–48, 53, 69; Hurricane Maria in Puerto Rico, 42, *68*, 69–73; Hurricane Sandy in New York, 60; North Atlantic basin (1850–2014), *71*; paths and strongest storms since 1979, *43*; typhoon Vera in Japan, 168; wind speeds and Category 5/Category 6 storms, 42, 69, 188n2. *See also* extratropical cyclones; storms, strengthening

"cytokine storms," 44

Daly, Herman, 64

Daphnia magna (zooplankton), 8

Darwin, Charles, 118

de Blasio, Bill, 62, 64

"dead zones" (oxygen minimum zones), 2, 14–19, *15*, *16*, 21; Arabian Sea, 14–19, *15*, 21; coastal, 2, *15*, *16*, 21

decolonization, 80–81, 97

de-densification, 154

deep adaptation, xii

deep ecology, xii, 139

deep time, xii

Deleuze, Gilles, 177

Delmarva Peninsula, 147

dengue fever, 87, 164

DePalma, Robert, 204n12

diatoms, 14, 106, *108*, 109

Dickens, Charles, 160–61

Digital Elevation Model (DEM) tools, *164*

dimethyl sulfide, 1

dinoflagellates, 14–17, 30, *108*, 116

dioxins, 38, 183n12

Dominica, 68, 69, 71

"dot tags," 12

dredging of shipping channels, 36, 52–53

DuPont, 36

Earth Institute at Columbia University, 62

East China Sea, 136, 138

eastern black rail, 145–47, 149, 202n5

eco-cities, 138, 140

eco-construction, 141

ecological economics, 64

ecology (definitions), 138–39

Economic Exclusion Zones (EEZs), 30

economy: balancing climate adaptation and economic development, 154–57; definitions, 64–65; environmental economics, 64–65; vulnerability of national economies to climate-caused changes in fisheries, *102*

ecosystems: definitions, 138–39; and resilience, 146

eco-tourism: Antarctica, 127, 128; coral reef tourism, 117, 118–21; Kenya, 117, 118–21

El Niño Southern Oscillation (ENSO), 87, *98*, 99–103, 116; and coral bleaching events, 116; sea surface temperatures and Peru's anchovy fishing industry, *98*, 99–103

Elbe River, *54*, 55, 58–59

Emmett, Ed, 46

Endangered Species List, 145, 147

endocrine-disrupting chemicals, 9

Englum, Lynn, 62

Enviromedics: The Impact of Climate Change on Human Health (Mery and Auerbach), 89

environmental economics, 64–65

Environmental Integrity Project, 48

environmental justice: children living within one mile of contamination sites, *52*; and climate change disasters, 43–44, 49–50; and downsides of nature-based solutions, 57; India's National Green Tribunal, 20;

and local land-use decisions, 49–50; low-income and minority residential neighborhoods, 46, *47*, 49–50, *52*

Environmental Justice Foundation, 79

Environmental Protection Agency (EPA), 36, 39

Environmental Working Group, 188n10

Ephesus, Turkey, 129

estuarine wetlands, 130, 142–49, *144*; adaptation, 144; anatomy of a temperate North American wetland, *144*; bird populations, 142–49; Chesapeake Bay, 142–49, *143*; salinity, 142–44; and sea-level rise, 130, 142–49, *143*, *148*; specialist species, 145–46; and tides, 142–44, *144*; wetlands mitigation, 144, 145

ethylene, 8, 48

European Chemical Agency, 10

European Union, 33, 81

eutrophication, 18

Evans, Jacqueline, 28, 30, 31–32, *33*

evolution: assisted, 117–18; definitions, 118

Extinction Rebellion, 158, 163

extratropical cyclones, 42

Exxon, 48, 65

fast-life ocean species, 101

Federal Emergency Management Agency (FEMA), 62, 64, 70

feedback loops, xiv, 89, 93–94, *95*, 111, 124; Arctic sea ice, 93–94, *94*, 95; biogeochemical, 1, *3*, 10, 18, 89, 106, 181n2; systems thinking, 111

fertilizers, nitrogen- and phosphorus-based, 2, 4, *15*, 18, 20–21

fever and immune systems, 88, 194n12

Feynman, Richard, 126

fish kills, 17

fishing industry, commercial: anchovy fishing of Peru, 99–105; and fast-life ocean species, 101; illegality and abuses in, 101–2, 103–5, 196n10, 197n14; industrial factory ships, 112; Kenya's small-scale fisheries, 117, 120–21; marine protected areas (MPAs), 30–32, 117, 121; North Atlantic cod fishery collapse, 110; Norway, 112; ocean currents and Southern Hemisphere fisheries, 127; overfishing, 17, 18, 19–20, 101; technological solutions to combat illegality in, 103–5; vulnerability of national economies to climate-caused changes, *102*

flood risk management, *51*, *57*; Hamburg's HafenCity and nature-based protection, 55–59; integrated, *57*; New York City's sea-level rise adaptation plans, 60–67

food choices, 27, 174; sustainable meat, 174; vegetarian

diets, 26, 27, 174

food insecurity, 25, 93; projected agricultural productivity (2050), *152*; rice crops and Mekong Delta flooding, *150*, 151–57

food waste, 25, *26*

FoodPrint.org, 174

foraminifera, 89, *108*, 110

Forest Stewardship Council, 172

France, 80–81

Francis, Pope, 170

frazil ("grease ice"), 90

Friends of the Earth Japan, 168

Fukakusa, Ayumi, 168

Fukushima nuclear disaster (2011), 169

Gaia Theory, 177

Gallant, Dennis, 25

Gambia, 80

Gambierdiscus (protist), 28–30, *30*

García Márquez, Gabriel, 99

Gates, Ruth, 114

glacial retreat, 122–28. *See also* Antarctica

Global Alliance for the Rights of Nature, 32

Global Compact for Safe, Orderly, and Regular Migration (GCM), 79, 80

global interfaith environmental movement, 167–68, 172; Religious Forestry Standards, 172; and Shinto leaders of Japan, 167–68, 172

global warming: IPCC projections, xiii, 5, 87, 96, 130, 138, 201n8; and land ice loss, 124; and oil companies, 65; reversing, 178; and sea ice melting, 90–94; and vulnerable UNESCO World Heritage Sites, *166*, *170*; and warming oceans, 87–89, 93; younger generations' concern and action, *40*

glyphosate, 183n12

Goethe, Johann Wolfgang von, 55

Goldman Environmental Prize, 31–32

Goodall, Jane, 142

Gore-Tex, 36

Gravesham Borough Council (Great Britain), 158

Great Barrier Reef Act of Australia, 30

Great Britain: carbon-reduction mileposts, 163; Extinction Rebellion activists, 158, 163; national climate strategy for sea-level rise, 162–65; *Thames Estuary 2100* plan, 161–62, 203n8, 203n9; Thames River and sea-level rise projections, 158–65. *See also* Thames River and Greater London area

Great Depression, 64

Great Pacific Garbage Patch (North Pacific Gyre), *6*, 9, 13, 87

Green Party (UK), 164–65

Green Revolution (1966–1985), 18, 157

"greenhouse gas effect," 95

greenhouse gas emissions (GHG): American household consumption, 174; atmospheric carbon dioxide levels, 88, 89, 96, 138, 201n8; carbon dioxide and ocean acidification, 24; cement-related emissions, *139*; clothing production, 8; during COVID-19 shutdowns (2020), 130, 200n4; and food waste, 25, *26*; GHG-free electricity alternatives, 174; HFCs, 4; IPCC projections, xiii, 5, 87, 96, 130, 138, 201n8; methane hydrate, 94, 96, 156; nitrous oxide, 17, *169*; possible future sea levels for different greenhouse gas pathways, *132*; and rice farming, 156; and transportation, 174

Greenland, *91*, 92, *107*

grief and transformation, 176

Grote Mandrenke ("Great Drowning of Men") (1392), 189n1

Guerin, Ayasha, 60

Gulf of Maine, *22*, 23–25

Gulf Stream, 85

Gulick, Esther, 187n2

gyres, 9, 87

HafenCity project (Hamburg, Germany), *54*, 55–59

Hafiz, Muhammad, 14

Hai Thach, 153

Haitham bin Tarik, Sultan of Oman, 19–20

Hamburg, Germany, *54*, 55–59; HafenCity project and nature-based flood protection, *54*, 55–59; livability scores, 56–57, 190n5; North Sea storms and flooding, 55–56, 58–59, 189n1

Hampton Roads, Virginia, 142, *143*, 146. *See also* Chesapeake Bay's estuarine wetlands

Hanson, Rick, 41

Haraway, Donna, 177

Hardin, Garrett, 103

Hardy, Alister, 198n17

Hasina, Sheikh, 78

Hawai'i, 7–13; coral mass bleaching event, 117; Kure Atoll and plastic pollution, *6*, 7–13

Hayhoe, Katherine, 177

health and human bodies, xiii–xiv, 2–3, 44–45, 88–89, 132–33; cancer, 9, 10, 133, 183n12, 200n9; chronic stress, 44–45; immune system, 44–45, 88–89, 133; inflammation and infections, 44, 88, 89, 133, 194n12;

microbiome, 88; PFAS exposure, 36–38, *38*, 188n10; plastics and health impacts, 9, 10–11; response to injuries and trauma, 44–45, 133; upset chemical balance and illness, 2–3

Helmlinger, Peter, 63

Herman, Judith, 69

Himalayan Plateau, 16, *136*

Honcho, Jinja, 167

Hospital for Tropical Diseases (London), 163–64

Houston, Texas, 46–53, *47*; Houston Ship Channel, 46–53, *47*, 188–89n2; Hurricane Harvey (2017), 42, 46–48, 53, 69; Manchester neighborhood and pollution, *47*, 49; petrochemical manufacturing and pollution, 46–49, *47*, 53, 188–89n2

Hu Jintao, 138

Huangpu River, 135–37

Human Rights Convention, 77

Humboldt Current, 85, 87, 99–100

Hurricane Alley, 48

Hurricane Dorian (2019), 188n2

Hurricane Harvey (2017), 42, 46–48, 53, 69

Hurricane Irma (2017), 69–70

Hurricane Maria (2017), 42, *68*, 69–73

Hurricane Sandy (2012) and post-Sandy recovery, 60–67

hurricanes. *See* cyclones, tropical (hurricanes or typhoons)

hydrofluorocarbons (HFCs), 4

ice. *See* land ice; sea ice

iceberg harvesting, 127

Imam, H. T., 78

"implicate order," theory of, 170

India: and Antarctic Treaty, 125; Arabian Sea dead zones, *15*, 16–18, 20–21; coastal zone management, 20–21

Indian Ocean, 117; gyre, 9, 87

Indigenous peoples of the Arctic, 92

individual action and the climate crisis: as citizens, 174–75; in communities, 175; as consumers, 174

Indonesia, *164*, 171

Indus River, 18

Industrial Revolution, 24

"innocent bystander effect," 44, 133

Insular Cases, 71

interfaith movement. *See* global interfaith environmental movement

Intergovernmental Panel on Climate Change (IPCC): formula for scientific uncertainty and confidence in projections, 126; greenhouse gas and global warming projections, xiii, 5, 87, 96, 130, 138, 201n8; on ocean acidification, 24; Oceans and Cryosphere report (2019), 86; on permafrost melting, 93–94; sea-level rise projections, 130, 138

Intergovernmental Science-Policy Platform on Biodiversity and Ecosystem Services, 145–46

internally displaced persons (IPDs), 76–77, *77*, 79

International Court of Justice at The Hague, 80, 193n9

International Criminal Court, 193n9

Inuit people of the Arctic, 92–93, 95–97, 195n4

Ise, Japan, *166*, 167–69, 172; Geku Shrine, *166*, 168, 172; Ise-Jingu Grand Shrine, *166*, 167–68, 204n3; Neku Shrine, 168, 172; sea-level rise and Shinto shrines, *166*, 167–69, 172; Uji Bridge, 167

Jacob, Klaus, 62–63

Jakarta, Indonesia, *164*

Japan: Buddhist temples, 168, business leaders and corporate-led climate planning, 170–71; coal-burning power plants, *169*, 171; "cradle-to-cradle" models of manufacturing, 11–12; environmental awareness in, 168–69; Koizumi's climate leadership, 171; net electricity generation by fuel (2000–2015), *169*; sea-level rise and Shinto shrines of Ise, *166*, 167–69, 172; Shinto leaders and global interfaith environmental movement, 167–68, 172; typhoon Vera (1959), 168

Jeandel, Catherine, 1

jellyfish, 17, *109*, 112–13, 198n27

Johnson, Nicholas, 122

Juhl, Andrew, 14

kelp, 20, 106

Kenya: coral collapse, 114–21; ecotourism, 117, 118–21; fisheries, 117, 120; Kenya Vision 2030 initiative, 117; Kisite-Mpunguti Marine National Park, 114, 117, 118–19; marine protected areas (MPAs), 117, 121

Kenyan Wildlife Service, 120

Kerr, Catherine, 187n2

key terms, xiii; adaptation, 26; climate migrant/climate refugee, 79; commons, 103; ecology, 138–39; economy, 64–65; evolution, 118; nature, 169–70; power, 72; precautionary principle, 19; resilience, 146; resources, 58; retreat, 154; rights, 32; risk, 50; system, 111; technology, 37–38; thresholds, 9–10; transilience, 162; uncertainty, 126; vulnerability, 94–95

keystone species, 114

Kien Tra-Mai, 154

Kigali Amendment (2017), 4

Kimmerer, Robin Wall, 1

king tides, 130–31

Kisite-Mpunguti Marine National Park (Kenya), 114, 117, 118–19

Koizumi, Junichiro, 171

Koizumi, Shinjiro, 171

Koslov, Liz, 62

Kriebel, David, 19

Kuhn, Thomas, 177, 206n3

Kure Atoll, Hawai'i, *6*, 7–13

Kurzweil, Ray, 37

Kutupalong Camp, Bangladesh, *74*, 75, 77–78, 80

Lamont-Doherty Laboratory at Columbia University, 14

land ice: Antarctica, 122–28, *123*, *126*; ice streams, *123*, 124; life cycle, 124; and sea-level rise, 122–28

land trusts, 148

landfills: London's Thames River region, 160; PFAS, 36; San Francisco Bay Area's Newby Island Recovery Park, 35–36

land-use decisions, local, 49–50, 62, 148

"last-chance capture" technologies, 12

Latour, Bruno, 165, 177

Laudato si' (Pope Francis), 170

leatherback turtles, 7–9, *10*

Liberal Party (UK), 165

Lindeman, Raymond L., 138

"Little Ice Age," 86

lobsters, American (*Homarus americanus*), 22–25

Lomonosov Ridge, 95

London. *See* Thames River and Greater London area

Lovelock, James, 177

magnesium sulfate, 141

malaria, 161, 163–64

Malaysia, 79

managed realignment, 154

manganese mining, *29*, 30–33

mangrove forests, 16, 76, 130, *148*

Mansten, Leo, 158, 164–65

Maori people of New Zealand, 186n11

Marae Moana protected area (Cook Islands), *29*, 30–32

marine bacteriology and virology, 112

marine protected areas (MPAs), *29*, 30–32, 117, 121

Martin, J. Andrew, 62

Marzeion, Ben, *166*

Massachusetts Institute of Technology (MIT), 50–53

Match-Mismatch Hypothesis, 198n13

McCluney, Fiona, 80

McLaughlin, Sylvia, 187n2

Mekong Delta of Vietnam, *150*, 151–57; agricultural policies, 152–57; climate migration, 153–54; droughts and floods, *150*, 151–53; land subsidence, 154; rice farming, *150*, 151–57, *156*; salt water intrusion on rice paddies, *150*, 153; sea-level rise projections, 153–54

Mekong Delta Plan, 155

Mekong River Commission, 153–54

Meridional Overturning Circulation (Ocean Conveyor Belt), 86–88, 127

Mery, Jay, 89

methane hydrate, 94, 96, 156

microbes, oil-eating, 50–53

microbial learning and memory, 112

microbiome, human, 88

microplastics, 7–9, 12

migration corridors, 148, 175

military bases and PFAS groundwater contamination, 37

Mill, John Stuart, 64

mining: Arctic oil/gas, 95, 96–97; Cook Islands' seabed and manganese nodules, *29*, 30–33

Mitchell, Alanna, 85

mitigation, climate. *See* climate mitigation

Monbiot, George, 158

monsoon rains and flooding, 76, 80, 151–53

monsoon winds, *15*, 16, 18

Montreal Protocol (1987), 4

Morton, Timothy, 7, 177

Mumbai, India, *15*, 17–18

Myanmar, *74*, 74–81; decreasing crop yields, 76, 77; extreme weather and strengthening storms in Rakhine State, *74*, 76–77; IPD camps, 76–77; Rakhine Buddhist majority, 76–77, 79; Rohingya persecution and refugees to Bangladesh, 75–81

Myanmar Times, 76

Naess, Arne, 139

nanoplastics, 8–9

National Aeronautics and Space Administration (NASA), 87, 94, 150

National Flood Insurance Program (NFIP), 63–64, 148

National Green Tribunal (India), 20–21

National Guard, 70, 191n4

National Microbiome Initiative, 112

National Oceanic and Atmospheric Administration (NOAA), xii, 42, 48, 116, 173

Nature, 125

nature (definitions), 169–70

nature-based solutions, 56–58

Netherlands, 155, 203n16

New York, New York, 60–67, *61*; Flexible Adaptation Pathways, 64; flood-prone areas, 60–62, *61*; Hurricane Sandy's storm surge (2012), 60, 190n2; land-use policies and building codes, 62; mandatory flood insurance, 63–64; Regional Coastal Commission, 66; Resiliency Office, 62; Resilient Edgemere plan, 62; rewilding projects, 66–67; sea-level rise adaptation plans, 60–67; storm barriers, 63

New York Times, 48, 70, 191n4

New Zealand: and Cook Islands' environmental protection, 32–33; legal personhood rights to the Whanguni River, 32, 33, 186n11

Newby Island Recovery Park (San Francisco Bay Area), 35–36

Next City, 57

Nguyen Duc, 151–54, 156–57

nitrogen loads, 18

nitrous oxide, 17, *169*

Noctiluca scintillans (dinoflagellate), 14–17, *15*, *17*

Non Self-Governing Territories (NGSTs), 81

North Atlantic ecosystem, 106–13; cod fishery collapse, 110; sea surface temperatures, *107*, 112; species migration and changing food web, 111

North Atlantic tropical storms (1850–2014), *71*

North Pacific Gyre ("Great Pacific Garbage Patch"), *6*, 9, 13, 87

North Pole, 85, 96

North Sea storms, 55–56, 58–59, 189n1

Northwest Passage, 95

Norway, 112, 171

Nunavut, *91*, 92–93, 95–97; the 1999 Land Claims Agreement, 97. *See also* Arctic Ocean warming

nutria (*Myocastor coypus*), 147–49

ocean acidification, 2, 23–25, *25*

ocean chemistry, xiv, 1–5; carbon dioxide absorption, 2; deoxygenation, 2, 5, 14–19, *15*, *16*; ocean acidification, 2, 23–25, *25*; ocean stratification, 5, 14–19, 87–88; oxygen minimum zones and "dead zones," 2, 14–19, *15*, *16*, 21; and pollution, 1–2

Ocean Conservancy, 104–5

Ocean Conveyor Belt, 86–88, 127

ocean currents, 85–88, *86*; El Niño Southern Oscillation (ENSO), 87, *98*, 98–103, 116; gyres, 9, 87; Humboldt Current, 85, 87, 99–100; Meridional Overturning

Circulation (Ocean Conveyor Belt), 86–88, 127; and Southern Hemisphere fisheries, 127

ocean deoxygenation, 2, *5*, 14–19, *15*, *16*

Ocean Foundation, 32–33

Ocean Memory Project, 112

ocean stratification, *5*, 14–19, 87–88

ocean warming, xiv, 2, 16–17, *22*, 25, *82–83*, 85–89; Arctic Ocean, 90–97; coastal Peru's fishing industry, 98–105; coral reef collapse, 114–21, *115*; and El Niño Southern Oscillation (ENSO), 87, *98*, 99–103, 116; jellyfish populations, *109*, 112–13; marine heatwaves, 116–17; North Atlantic ecosystem and microscopic marine life, 106–13; and ocean currents, 85–88, *86*; and ocean stratification, *5*, 14–19, 87–88; Pine Island Glacier in Antarctica, 122–28; rates and projections, *82–83*, 87, 88–89; upwelling, 86–87

oil companies, 65

oil spills, 52–53, 96–97, 196n16; remediation, 52–53

oil-eating microbes, 50–53

Oman, *15*, *16*, 19–20

One Planet Sovereign Wealth Fund Working Group, 171

organochlorine pesticides (OCPs), 188n10

Ostrom, Elinor, 103

overfishing: Arabian Sea, 17, 18, 19–20, and fast-life ocean species, 101

oxygen minimum zones (OMZs) ("dead zones"), 2, 14–19, *15*, *16*, 21

Pakistan, *15*, *16*, 18, 20–21

Paleocene-Eocene Thermal Maximum, 89

Palmer, Martin, 167, 172

Palumbi, Stephen, *115*

Papahānaumokuākea National Marine Sanctuary (Hawai'i), *6*, 7

paradigm shifts, 177, 206n3

Paris Climate Accord (2015), xiii, 5, 168

Particularly Sensitive Sea Areas, 7

Pearl River Delta, China, *164*

People's Committee of Bến Tre (Vietnam), 153

perfluorinated compounds (PFCs), 188n10

perfluorooctanoic acid (PFOA), *38*

permafrost, 93–94, 96

personhood, legal, 31–33, 186n11

Peru, coastal, 89, *98*, 99–105; ENSO patterns, *98*, 99–103; Exclusive Economic Zone, 100; government efforts to increase *anchoveta* consumption, 100, 196n5; Peruvian *anchoveta* and fishing industry, 99–103, *101*;

Pisco, 99–101, 105; sea surface temperature anomalies during extreme ENSO years, *98*

petrochemical manufacturing, 36, 46–49, 188–89n2; Houston, 46–49, *47*, 53, 188–89n2; plastics production, 10–12

PFAS (per- and polyfluoroalkyl substances), 9, 36–39, *38*; blood levels in exposed groups, 36, 38, *38*, 188n10; groundwater contamination, 36–37, 187n7; health effects, 36–38, 188n10

phthalates, 9

phytoplankton, 16–20, 28–30, 87, 88, 106–8, *108*; cyanobacteria, 106–8, 197n7; diatoms, 14, 106, *108*, 109; protists, 16–17, 28–30, 88, 106, 108–9, 110

Pine Island Glacier (PIG) in Antarctica, 122–28, *123*

plankton, 1, 8, 18, 24, 100, 106–13; cyanobacteria, 106–8, 197n7; dinoflagellates, 14–17, 30, *108*, 116; and North Atlantic sea surface temperatures, *107*, 112; phytoplankton, 16–20, 28–30, 87, 88, 106–8, *108*; zooplankton, 8, 14, 24, 109–10

plastic bottles, 3–4, 8, 174

plastic production, 2, 6–13; clothing manufacturing, 7–9, 12–13, 184n22; global production and future trends, *9*, 13, 183n21; health impacts, 8–9, 10–11, 12; microplastics, 7–9, 12; North Pacific Gyre ("Great Pacific Garbage Patch"), *6*, 9, 13; pollution, 2, *6*, 6–13; polymers, 8, 10, 11; recycling, 9, 11–12, 183n17; reducing household consumption and alternatives, 174; single-use plastic, 3, 11

Plastics Industry Association, 11

Pleasanton, California, 187n7

Point Blue, 146

Pollan, Michael, 23

polybrominated diphenyl ethers (PBDEs), 188n10

polychlorinated biphenals (PCBs), 9, 188n10

polyester, 8, 12, 182n6

polyethylene beads, 7–8

polyethylene terephthalate (PET), 8

Post Carbon Institute, 146

post-traumatic stress disorder and growth, 45

power (definitions), 72

precautionary principle (definitions), 19

Prochlorococcus (cyanobacteria), 108

"producer pays" taxes, 11

protists, 16–17, 28–30, 88, 106, 108–9, 110

public housing, 49–50

public transportation, 174

Puerto Rican Department of Public Safety, 70

Puerto Rico, *68*, 69–73; diaspora and emigration to U.S.

mainland, 71–73; federal disaster response, 69–72, 191n4; Hurricane Maria (2017), 42, *68*, 69–73; San Juan's La Perla district, 73; statehood supporters, 71–72; territorial status and lack of sovereignty, 70–72

Puna, Henry, 30, 31

Queste, Bastien, 17, 184n6

REACH program regulations (European Commission), 10

Rebuild by Design, 62

recycling, plastic, 9, 11–12, 183n17

reforesting the sea, 20, 88

regime shift, *16*, 18–19

Regional Organization for Protection of the Marine Environment (ROPME), 185n15

Religious Forestry Standards, 172

remediation, post-natural disaster, 49

resilience: definitions, 146; New York City's climate adaptation plans, 62; and transilience, 162; U.S. Climate Resilience Toolkit, 138–39

resources (definitions), 58

retreat: definitions, 154; and sea-level rise, 154, 203n15. *See also* climate migrants

rice farming: Mekong Delta, *150*, 151–57, *156*; Myanmar, 76

Rice University, 48, 49

rights: definitions, 32; legal personhood, 31–33

rights-of-nature movement, 31–33

Rilke, Rainer Maria, 176

risk (definitions), 50. *See also* climate risk management

Robinson, Kim Stanley, 173

Romanticism, 169

Rosenzweig, Cynthia, *150*

Rush, Elizabeth, 144

Russia: and Antarctic Treaty, 125; Russian Arctic, 92, 95, 96

Salk, Jonas, 178

salt marshes. *See* estuarine wetlands

saltwater intrusion, 76, 135, *150*, 153; estuarine wetlands, 142–44; Mekong Delta of Vietnam, *150*, 153; and rice crops, 76, *150*, 153

San Francisco Bay Area, xii, *34*, 34–40; "Do Not Eat" list on fishing piers, *37*; Newby Island Recovery Park (landfill), 35–36; toxics pollution, *34*, 35–37

San Jose, California, 36

San Juan, Puerto Rico, 69–73

Santoro, Alyson, 110

Saudi Public Investment Fund, 171

Schmidt, Gavin, 94

scientific method, 125, 126

Scotchgard, 36

Scotland, 204n15

Scranton, Roy, 48

sea butterflies, 24

sea ice: Antarctic, 125; Arctic Ocean, 90–97, *91*, *93*, *94*; feedback loops, 93–94, *94*, 95; first-year, 90, *91*, *93*; frazil ("grease ice"), 90; and global warming, 90–94; ice-albedo feedback and Arctic amplification, 93, *94*; multi-year, 90, *91*, *93*; pancake ice, 90

Sea Shepherd, 104–5

Seabed Minerals Act of 2019 (Cook Islands), 32

seabed mining, *29*, 30–33

seabirds, 7, 102, 117

seafood poisoning, non-bacterial, 28–30

SeafoodWatch.org, 174, 196n10

sea-level rise, xiv, 129–33; and Antarctic meltwater, 122–28; contributors to (1993–2018), *131*; estuarine wetlands, 130, 142–49, *143*, *148*; impacts on agricultural land, *155*; IPCC projections, 130, 138; land ice loss, 122–28, *126*; land subsidence, 131, *134*, 135–38, *137*, 140, 141, 154; long-range planning and policy decisions for, 131–32; management through adaptation, 144; management through mitigation, 144, 145; migration and retreat from, *78*, 153–54, 203n15; projected global mean sea-level change to 2300, *160*; projected sea levels for different greenhouse gas pathways, *132*; projections based on SRTM data vs. DEM topography, *164*; "slow emergency" of climate change, 41; and thermal expansion, 130, *131*; and tidal cycle, 130–31, 142–44; and UNESCO World Heritage Sites, *166*, *170*. *See also* sea-level rise and coastal cities

sea-level rise and coastal cities, 60–67, 129, *164*; Chesapeake Bay region, 142–49, *143*; land subsidence, 131, *134*, 135–38, *137*, 140, 141, 154; London and the Thames River estuary, 158–65, *159*; Mekong Delta flooding, *150*, 150–57; New York adaptation and planning, 60–67; San Francisco Bay area, *34*; Shanghai, *134*, 135–41; Shinto shrines of Japan, *166*, 167–69, 172

Seed to Table (Japan), 156

Semonite, Todd, 70

Serres, Michel, 169–70, 177

Servan-Schreiber, David, 200n9

Shanghai, China, *134*, 135–41; eco-cities, 138, 140–41;

flooding, 136–38, 139–41; freshwater aquifer, 135–36; Hengsha Island, *134*, 138; land subsidence, *134*, 135–38, 140, 141; Nanhui New City, 138; Pudong District, 135, 140; seawalls and dikes, 137–38, 140

shark species, endangered, 17, 20, *20*

Shell, 65

Shinto religion of Japan, 167–69, 172; and global interfaith environmental movement, 167–68, 172; *kami*, 168, 172; Shinto shrines, *166*, 167–69, 172, 205n8

Shuttle Radar Topography (SRTM), *164*

Silicon Valley, 35, 37

Sinopec refineries, 8

Skara Brae, Scotland, 129

Sloterdijk, Pieter, 177

Social Democratic Party (Germany), 59

Solnit, Rebecca, 35

Sosik, Heidi, 110

South Africa, 127

South Asian Sea Program (SASP), 185n15

South Pacific Gyre, 87

South Pacific island nations, 33

specialist species, 145–46

Spilhaus, Athelstan, *xvii*

Spilhaus Projection map, *xvi–xvii*

spotted eagle ray (*Aetobatus narinari*), *18*

Staten Island, *61*, 62, 66

steady-state economics, 64

Stengers, Isabelle, 177

Stockholm Resilience Center's "Nine Planetary Boundaries," *3*, 10, 18

storms, strengthening, xiv, 41–45; and armed conflict, 76–77; billion-dollar of storm and flood disaster events (U.S.), *63*; coastal urban population centers, 42–43, 45, 60, *66*, 140; emergency preparation, disaster responses, and recovery, 43–45; extratropical cyclones, 42; flood risk management, *51*; highest one-day precipitation events (1910–2015), *48*; inequalities in storm response and recovery, 43–44, 49–50; Myanmar, *74*, 76–77; North Sea, 55–56, 58–59, 189n1; storm frequency models, 42; storm intensity models, 42, 46–48; thunderstorms, 42; tropical cyclones, 42, *43*, 46–48, 68, *71*, 140. *See also* cyclones, tropical (hurricanes or typhoons)

Strauss, Ben, 129

stress, chronic, 44–45

The Structure of Scientific Revolutions (Kuhn), 177

Suga, Yoshihide, 171

Sundarbans, 130, *164*

Superfund sites, *47*, 49, *52*

Surfrider Foundation, 104–5

survival migration, 79

sustainability, 203n15

Svalbard (Arctic archipelago), 33, *91*, 96

system (definitions), 111

systems theory, 111

syzygy, 130

Takigawa, Christel, 171

Tang Yiqun, 137

Tansley, A. G., 138

Task Force on Climate-Related Financial Disclosures (TCFD), 170–71

technology (definitions), 37–38

Teflon, 36

Ten Principles of Jurisprudence, 32

Texas A&M University, 48

Texas Tech University, 177

Thames Estuary 2100 (*TE 2100*), 161–62, 203n8, 203n9

Thames Estuary Growth Commission, 161

Thames River and Greater London area, 158–65, *159*; floods, *159*, 160; malaria and waterborne diseases, 161, 163–64; Rainham Marshes and Landfill, 160; River Roding, 160; Royal Docks neighborhood, 160; sea-level rise and projections, 158–65, *159*; seawalls, dikes, and barriers, 158–62, *159*, 165; *TE 2100* plan, 161–62, 203n8, 203n9; Thames Barrier, 158–60, *159*; Thames Estuary, 158–65, *159*; Thames River seals, *162*; tides, 160; towns slated for abandonment and defense, 161–62, 204n11

Thengar Char refugee camp (Bangladesh), 77–78

3M, 36

thresholds (definitions), 9–10

thunderstorms, 42

Tickner, Joel, 19

tides: estuarine wetlands, 142–44, *144*; king tides, 130–31; and sea-level rise, 130–31, 142–44; Thames River and Greater London area, 160

tiger shrimp farms, 153

"tipping points," ecological, 9–10

tomopteris, 110

Tongji University (Shanghai), 137

toxic algal blooms, 2, *31*, 33

Toxic Substances Control Act (TSCA), 39

toxics pollution: PFAS, 36–39, *38*, 187n7; San Francisco Bay Area, *34*, 35–37

"tragedy of the commons," 103

Tran Thuc, 153–54

transilience, 162, 176–78, 204n12

Treaty of Waitangi (New Zealand), 32

Tropical Depression Imelda (2019), 46

Trump, Donald, 36, 69

tuna, 102, 196n10

Turner, Kelly, 133

typhoons. *See* cyclones, tropical (hurricanes or typhoons)

Uganda, 80

Ulva (green macroalgae), xii

uncertainty (definitions), 126

"Undersea" (Carson), 197n3

UNESCO World Heritage sites, 7, *166*, *170*

United Nations: clean-tech program, 12; and climate migrants, 80–81; and eradication of colonialism, 80–81; global biodiversity stability plans, 30

United Nations Convention on International Trade in Endangered Species (CITES), 105

United Nations Development Program, 80

United Nations Food and Agriculture Organization (FAO), 102

United Nations Human Rights Council (HRC), 79

United Nations Intergovernmental Science-Policy Platform on Biodiversity and Ecosystem Services, 145–46

United Nations Special Envoy on Migration, 80

United States: and Antarctic Treaty, 125; billion-dollar of storm and flood disaster events (1980–2020), *63*; food waste, 25; long-range planning for sea-level rise, 131–32; PFAS production ban, 36; plastics consumption and pollution, *6*; regulation and climate change policies, 5–6, 10

University of California at Los Angeles, 62

University of California at San Francisco, 188n10

University of California at Santa Barbara, 110

University of Maine, 25

University of Texas, 48

Urbina, Ian, 197n14

U.S. Climate Resilience Toolkit, 138–39

U.S. Department of Defense, 70

U.S. Fish and Wildlife Service (USFWS), 145, 147, 148–49, 197n3

U.S. Geological Survey (USGS), xii

Vallely, Paul, 167

Vanuatu, 129

vegetarian diets, 26, 27, 174

Vernadsky, Vladimir, 181n2

Verrazano Narrows (New York City), 63

Vietnam: agricultural policy, 152; balancing economic development and climate adaptation, 154–57; Communist Party, 152, 154–55; Mekong Delta flooding, *150*, 150–57. *See also* Mekong Delta of Vietnam

Vietnam Institute of Meteorology, Hydrology, and Environment, 153–54

Vietnam War, 152

viruses, marine, 110–11

vulnerability (definitions), 94–95

Wadhams, Peter, 93, 94

Wales, 204n11

war and climate change, 76–77

warming waters. *See* ocean warming

waterborne diseases, 161, 163–64

West Antarctic Ice Sheet, 124–25, 127

wetlands mitigation, 144, 145

Wetlands Oversight Boards, 147

Whyte, Kyle, 90

Wiener, Norbert, 111

Woods Hole Oceanographic Institution, 110

World Bank, 71, *77*

World Health Organization, 19, 25

World War II, 4, 79, 92

WorldCare Standard, 172

Worldwide Fund for Nature, 172

Wright, Beverly, 49–50

Xi Jinping, 138

Yancheng, China, 137

Yangtze Delta Region, 12, *134*, 135–41, *136*

Yokosuka, Japan, 171

young people: Alphas (children of Millennials), 38–40; climate activists, 158; climate change attitudes of Japanese youth, 168–69; climate change concern and action, *40*, 168–69

Youth Climate Strike in Tokyo (September 2019), 168

Zarrilli, Daniel, 64, 67

zero-emission communities, 140–41

Zhang Yue, 141

zooplankton, 8, 14, 24, *108*, 109–10, 116

zooxanthellae (dinoflagellate), 116

CHRISTINA CONKLIN is an artist, writer, and researcher whose work investigates impermanence and possibility, often using the ocean as both site and metaphor. Her essays, exhibitions, events, and interactive installations consider our personal and societal responses to the intersecting global crises of our time. She is currently working with thought leaders and activists to grow emergent, regenerative systems in the San Francisco Bay Area and beyond.

MARINA PSAROS is a sustainability expert and has led climate action programs across public, private, and nonprofit organizations for over a decade. She is one of the creators of the King Tides Project, an international community science and education initiative. An amateur cartographer and ocean advocate, she lives in the San Francisco Bay Area.

PUBLISHING IN
THE PUBLIC INTEREST

Thank you for reading this book published by The New Press. The New Press is a nonprofit, public interest publisher. New Press books and authors play a crucial role in sparking conversations about the key political and social issues of our day.

We hope you enjoyed this book and that you will stay in touch with The New Press. Here are a few ways to stay up to date with our books, events, and the issues we cover:

Sign up at www.thenewpress.com/subscribe to receive updates on New Press authors and issues and to be notified about local events

Like us on Facebook:
www.facebook.com/newpressbooks

Follow us on Twitter:
www.twitter.com/thenewpress

Follow us on Instagram:
www.instagram.com/thenewpress

Please consider buying New Press books for yourself; for friends and family; or to donate to schools, libraries, community centers, prison libraries, and other organizations involved with the issues our authors write about. The New Press is a 501(c)(3) nonprofit organization. You can also support our work with a tax-deductible gift by visiting www.thenewpress.com /donate.